COMPLETE GUIDE
to VHS
CAMCORDER
TROUBLESHOOTING
and REPAIR

Other books by John D. Lenk:

Complete Guide to Compact Disc (CD) Player Troubleshooting and Repair—1986

Complete Guide to Laser/Videodisc Player Troubleshooting and Repair—1985

Complete Guide to Modern VCR Troubleshooting and Repair—1985

Complete Guide to Telephone Equipment Troubleshooting and Repair—1987

Complete Guide to Videocassette Recorder Operation and Servicing—1983

COMPLETE GUIDE to VHS CAMCORDER TROUBLESHOOTING and REPAIR

JOHN D. LENK
Consulting Technical Writer

Prentice Hall, Englewood Cliffs, New Jersey 07632

Library of Congress Cataloging-in-Publication Data

LENK, JOHN D.
  Complete guide to VHS camcorder troubleshooting and repair.

  Includes index.
  1. Camcorders—Maintenance and repair. I. Title.
II. Title: VHS camcorder troubleshooting and repair.
TR882.L46 1988   621.388′3   87-14545
ISBN 0-13-160862-2

© 1988 by Prentice-Hall, Inc.
A Division of Simon & Schuster
Englewood Cliffs, New Jersey 07632

The publisher offers discounts on this book when ordered
in bulk quantities. For more information, write:

  Special Sales/College Marketing
  Prentice-Hall, Inc.
  College Technical and Reference Division
  Englewood Cliffs, NJ 07632

*All rights reserved. No part of this book may be
reproduced, in any form or by any means,
without permission in writing from the publisher.*

Printed in the United States of America

10 9 8 7 6 5 4 3 2 1

ISBN 0-13-160862-2

PRENTICE-HALL INTERNATIONAL (UK) LIMITED, *London*
PRENTICE-HALL OF AUSTRALIA PTY. LIMITED, *Sydney*
PRENTICE-HALL CANADA INC., *Toronto*
PRENTICE-HALL HISPANOAMERICANA, S.A., *Mexico*
PRENTICE-HALL OF INDIA PRIVATE LIMITED, *New Delhi*
PRENTICE-HALL OF JAPAN, INC., *Tokyo*
SIMON & SCHUSTER ASIA PTE. LTD., *Singapore*
EDITORA PRENTICE-HALL DO BRASIL, LTDA., *Rio de Janeiro*

This book is dedicated to my wife Irene,
whose encouragement has made the book possible, and whose patience has made the work bearable.
The book is also dedicated to Karen, Tom,
Brandon, and Justin.
And to our very special Lambie, be happy always.
You are a very good boy!
And to all who read this book
Champagne Wishes
Caviar Dreams!

CONTENTS

Preface *xi*

1 Introduction to Camcorders *1*

1-1 The Basic Camcorder *1*
1-2 Comparison of Camcorder and Standard VHS Mechanisms *5*
1-3 Color Camera Basics *9*
1-4 Typical Camcorder Features *19*
1-5 Typical Camcorder Specifications *23*

2 User Controls, Operation Procedures, Accessories, and Interconnections *25*

2-1 Accessories *25*
2-2 Typical Operating Controls and Indicators *27*
2-3 Interconnections *36*
2-4 Typical Operating Procedures *43*
2-5 Precautions *56*

3 Test Equipment, Tools, and Routine Maintenance 58

- 3-1 Safety Precautions in Camcorder Service 58
- 3-2 Test Equipment for Camcorder Service 59
- 3-3 Tools and Fixtures for Camcorder Service 63
- 3-4 Routine Maintenance for Camcorders 65

4 Camera Circuits 68

- 4-1 Camera Basics 68
- 4-2 Camera Power Distribution Circuits 74
- 4-3 Camera Deflection Circuits 77
- 4-4 Camera Signal-Processing Circuits (with Pickup Tube) 90
- 4-5 Miscellaneous Camera Circuits 112
- 4-6 Camera Signal-Processing Circuits (with MOS Pickup) 118

5 Tape Transport and Servo System 141

- 5-1 Tape Transport Basics 141
- 5-2 Mechanical Operation of Typical VHS Camcorder 142
- 5-3 Servo System Basics 151
- 5-4 Cylinder Control 154
- 5-5 Capstan Control 166
- 5-6 CTL Pulse Recording/Playback Path 172
- 5-7 Artificial V-Pulse Generator 173
- 5-8 Phase Matching 175

6 Video/Audio Signal-Processing Circuits 177

- 6-1 Video/Audio Signal-Processing Basics 177
- 6-2 Video In/Out-Selection Circuits 178
- 6-3 Luma/Chroma Record Process 180
- 6-4 Luma/Chroma Playback Process 182
- 6-5 Video Headswitching Signal Generation 185
- 6-6 Video Head-Select Operation 187

Contents

6–7 Video Head Playback-Select Operation *190*
6–8 Audio Signal-Processing Circuits *192*
6–9 Luminance (Black and White) Troubleshooting *197*
6–10 Chroma Troubleshooting *199*

7 System Control and Electrical Distribution *201*

7–1 Electrical Overview *202*
7–2 AC Adapter/Charger *203*
7–3 System-Control Overview *206*
7–4 Power-Control Circuits *207*
7–5 Battery Overdischarge Detection Circuits *211*
7–6 Function Switch Circuits *215*
7–7 Trouble Sensor Circuits *217*
7–8 Tape Counter Circuits *220*
7–9 Mode Indicator Circuits *223*
7–10 Loading Motor Circuits *225*
7–11 Capstan Motor Circuits *227*
7–12 Mode Control Circuits *230*

8 Electrical and Mechanical Adjustments *234*

8–1 Mechanical Adjustments *235*
8–2 Electrical Adjustments *247*

9 Troubleshooting and Service Notes *284*

9–1 The Basic Troubleshooting Approach *284*
9–2 Operational Checklist *287*
9–3 Camcorder Troubleshooting/Repair Notes *288*
9–4 Trouble Symptoms Related to Adjustment *293*
9–5 Using the Troubleshooting Procedures *295*
9–6 No Picture on EVF or Monitor Power On, Camera Mode *295*
9–7 No Record Video *296*
9–8 No Playback Video *296*
9–9 No Power to Camera with EVF Raster On *297*
9–10 No-Record Chroma *298*

- 9-11 No Playback Chroma *298*
- 9-12 Power Zoom Does Not Operate *298*
- 9-13 Video Level Too High or Low *299*
- 9-14 No Picture *299*
- 9-15 No Autofocus *300*
- 9-16 No Color *301*
- 9-17 Incorrect Color Shading *302*
- 9-18 Incorrect White Balance (Incorrect Color Balance) *303*
- 9-19 No EVF Raster *304*
- 9-20 No EVF Picture *304*
- 9-21 No Sound (Audio) *305*
- 9-22 No Record Audio *306*
- 9-23 No Playback Audio *306*
- 9-24 No Power *307*
- 9-25 Battery Overdischarge is Not Detected *308*
- 9-26 Function Switches Inoperative or Malfunctioning *308*
- 9-27 Trouble Detection Circuits Inoperative or Malfunctioning *309*
- 9-28 Tape Counter Inoperative *309*
- 9-29 Cassette Does Not Eject *309*
- 9-30 Mode Indicator Displays Inoperative or Malfunctioning *309*
- 9-31 Horizontal Stripes on Display *310*
- 9-32 Cylinder Does Not Rotate *310*
- 9-33 Capstan Does Not Rotate *311*
- 9-34 Picture Swings Horizontally, or Noisy Picture and Clean Picture Alternately Appear on the Display in Playback *311*
- 9-35 Picture Swings Vertically, or Noisy Picture and Clean Picture Alternately Appear on the Display in Playback *312*
- 9-36 Many Noises Appear on Screen in Playback Mode *313*
- 9-37 Noisy Picture and Clean Picture Alternately Appear on Screen (Only When Prerecorded Tape is Played Back) *313*

Index *315*

PREFACE

The purpose of this book is to provide a simplified, practical system of troubleshooting and repair for the many types and models of VHS camcorders (camera-recorders). The book is the ideal companion to the author's *Complete Guide to Videocassette Recorder Operation and Servicing*[1] and *Complete Guide to Modern VCR Troubleshooting and Repair.*[2] However, there is no reference to either book, nor is it necessary to have any other book to make full use of the information presented here.

Of course, it is assumed that you are already familiar with the basics of television and magnetic recording, and with VHS VCRs. If not, do not attempt troubleshooting and repair of any camcorder (at least not until you have read this book!). Camcorders are probably the most complex devices found in consumer electronics. Even a basic camcorder includes all the components found in a VCR (with which some technicans are familiar), plus the components of a full-feature color video camera (with which most technicians are not familiar).

You need schematic diagrams, part-location photos or drawings, descriptions of adjustment procedures, and so on to do a proper troubleshooting/repair job. Very simply, you must have adequate service literature for any specific model of camcorder that you are servicing. Instead of trying to duplicate such details, this book concentrates on troubleshooting/repair *approaches*. This is done in several ways.

[1]Prentice-Hall, Inc., Englewood Cliffs, N.J., 1983.
[2]Prentice-Hall, Inc., Englewood Cliffs, N.J., 1985.

First, most service literature gives you the adjustment procedures for a given model of equipment and shows the physical location of the adjustment points for that model, but does not tell what is being done by the adjustment. In this book, we start by describing the *purpose of the adjustment* and then make reference to diagrams that show the *electrical position* (or circuit location) of input/output, adjustment controls, and so on.

Likewise, the service literature often does not include the "how it works" or theory of operation (on the assumption that everyone knows how a camcorder works). In this book, we provide *full circuit-by-circuit theory* for all sections of the camcorder.

To simplify the "how it works" discussions, a typical camcorder is broken down into its various circuits or sections. Camcorders of all manufacturers have certain circuits and/or sections in common (camera section, tape transport, servo, signal-processing circuits, and system control). A separate chapter is devoted to each of the major sections.

Using this chapter/circuit-group approach, you can quickly locate information you need to troubleshoot a malfunctioning camcorder. In each theory chapter, you will find (1) an introduction that describes the purpose or function of the circuit, (2) some typical circuit descriptions or circuit theory (drawn from a cross section of camcorders), and (3) a logical troubleshooting/repair approach for the circuit (based on manufacturers' recommendations).

Finally, if the troubleshooting approach found in the theory chapter does not locate a specific problem, you can refer to the troubleshooting chapter. This final chapter describes troubleshooting from the standpoint of *failure* or *trouble symptoms* (the most common trouble symptoms reported to manufacturers' service personnel). Each trouble symptom is related directly to a specific circuit, adjustment control, or group of circuits that might logically produce the symptom.

Chapter 1 is devoted to the basics of camcorders. This includes technical specifications and the relationship of VHS camcorders to standard VHS VCRs. The chapter also introduces color video cameras and color-processing principles.

Chapter 2 describes user controls, operating procedures, accessories, and typical interconnections between the camcorder and TV or VCRs. The chapter concentrates on the functions of operating controls (to establish a basis for troubleshooting) rather than on camera techniques.

Chapter 3 describes the test equipment and tools needed for camcorder service. The chapter also discusses routine maintenance for camcorders, including lubrication and cleaning.

Chapter 4 describes the theory of operation for a cross section of camera circuits, including those with pickup tubes, and late-model cameras with solid-state image sensors. By studying the circuits found here, you should have no difficulty understanding the schematics and block diagrams of similar camcorders.

Circuit descriptions are supplemented with partial schematics and block diagrams that show such important areas as signal flow paths, input/output, adjustment controls, test points, and power source connections (the areas most important for troubleshooting).

The chapter concludes with a discussion of the troubleshooting approach for camera circuits.

Chapter 5 describes the theory of operation and basic troubleshooting approach for a cross section of tape-transport and servo systems, using the same level and format found in Chapter 4.

Chapter 6 describes the theory of operation and basic troubleshooting approach for signal-processing circuits, using the same level and format found in Chapter 4.

Chapter 7 describes the theory of operation and basic troubleshooting approach for system control and electrical distribution, using the same level and format found in Chapter 4.

Chapter 8 describes adjustment procedures, both mechanical and electrical, for a typical VHS camcorder. The procedures are included to show what is involved for complete adjustment/alignment and are not intended to replace procedures for a specific camcorder.

Using these examples, you should be able to relate the procedures to a similar set of adjustments on most camcorders. Likewise, by studying the test points, signal voltages, and waveforms given here, you should be able to identify corresponding waveforms and voltages on the camcorder you are servicing (even though the signals may appear at different points).

Chapter 9 describes troubleshooting and service notes for a cross-section of VHS camcorders. The chapter starts with a review of the author's basic troubleshooting approach and then goes on to circuit-by-circuit troubleshooting.

This step-by-step approach is based on failure or trouble symptoms and represents the combined experience and knowledge of many camcorder service specialists and service managers.

Many professionals have contributed their talent and knowledge to the writing of this book. The author gratefully acknowledges that the tremendous effort to make this book such a comprehensive work is impossible for one person and wishes to thank all who have contributed, directly and indirectly.

The author wishes to give special thanks to the Publication Division of Hitachi Sales Corporation of America, Deborah Fee and Pat Wilson of N.A.P. Consumer Electronics (Magnavox, Sylvania, Philco), Thomas Lauterback of Quasar, Judith Fleming and J. W. Phipps of GE/RCA Consumer Electronics, and John Taylor and Matthew Mirapaul of Zenith.

The author extends his gratitude to Paul Becker, Gloria Colagreco, Elma Simeone, Matt Fox, Diane Spina, Melissa Halverstadt, Greg Burnell, Hank Kennedy, John Davis, Jerry Slawney, Irene Springer, Barbara Cassel, Karen Fortgang, Lisa Schulz, Erica Orloff, Natalie Brenner, Judy Winthrop,

Pat Walsh, Ellen Denning, Linda Maxwell, Jewel Harris, Armond Fangschlyster, and Rudy Drelb of Prentice-Hall and Ann Marie Norris of Prentice-Hall International. Their faith in the author has given him encouragement, and their editorial/marketing expertise has made many of the author's books international best sellers. The credit must go to them.

The author also wishes to thank Joseph A. Labok of Los Angeles Valley College for his help and encouragement throughout the years.

And to my wife, Irene Lenk, my research analyst, I wish to thank her. Without her help, this book could not have been written.

JOHN D. LENK

**COMPLETE GUIDE
to VHS
CAMCORDER
TROUBLESHOOTING
and REPAIR**

1

INTRODUCTION to CAMCORDERS

This chapter is devoted to the basics of VHS camcorders (camera-recorders). A VHS camcorder is a combination color video camera and VCR (video-cassette recorder) using standard VHS cassette tape. Camcorders contain conventional video-camera/VCR signal-processing circuits and single-speed (SP-mode) VCR mechanics. However, the physical size of the VCR mechanism is greatly reduced from that found in conventional VCRs. The major difference between the camcorder mechanism and conventional VCRs is a reduction in size of the cylinder or "scanner."

1-1 THE BASIC CAMCORDER

Figures 1-1a through 1-1f show some typical VHS camcorders. Figure 1-2 is a simplified block diagram of a typical or generalized camcorder.

As shown in Figs. 1-1 and 1-2, a camcorder combines the convenience and features of a deluxe video camera with those of a VHS VCR. There is no need for cable connections to a portable VCR. Likewise, the camcorder provides up to 2 hr and 40 min of recording with a standard T-160 VHS cassette. (However, with most VHS camcorders, it is necessary to use an a-c adapter/charger or an optional battery pack to record for more than 1 hr.)

Before we get into the features of a typical VHS camcorder, let us review some of the basics, such as differences between the camcorder and standard VHS mechanics and the principles of color cameras.

FIGURE 1-1a RCA CMR200 ProWonder™ Camcorder (Courtesy of GE/RCA Consumer Electronics)

FIGURE 1-1b RCA CMR300 ProWonder™ Camcorder (Courtesy of GE/RCA Consumer Electronics)

Sec. 1-1 The Basic Camcorder 3

FIGURE 1-1c Quasar VM-11 VHS Movie Camcorder (Courtesy of Quasar Company, Elk Grove Village, Illinois)

FIGURE 1-1d Sylvania VCC151 VHS Movie System (Courtesy of N.A.P. Consumer Electronics Corporation–Sylvania Audio-Video Product)

FIGURE 1-1e Hitachi VM5000A VHS Movie Camcorder (Courtesy of Hitachi Sales Corporation of America)

FIGURE 1-1f Zenith VM7100 VHS Video Camera/Recorder (Courtesy of Zenith Electronics Corporation)

Sec. 1-2   Comparison of Camcorder and Standard VHS Mechanisms   5

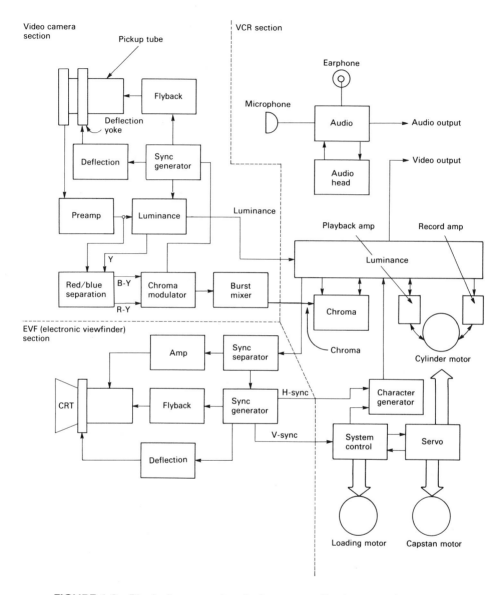

FIGURE 1-2   Block diagram of typical or generalized camcorder

1-2 COMPARISON OF CAMCORDER AND STANDARD VHS MECHANISMS

Figure 1-3 shows the basic mechanical difference between a camcorder and a VHS VCR. With the standard 2-head VCR, the tape is wrapped approximately 180° around a 62-mm cylinder that rotates at 1800 rpm. With a VHS

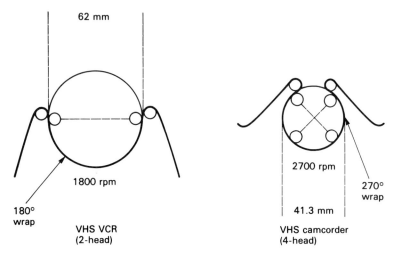

FIGURE 1-3 Basic mechanical difference between a VHS camcorder and a VHS VCR

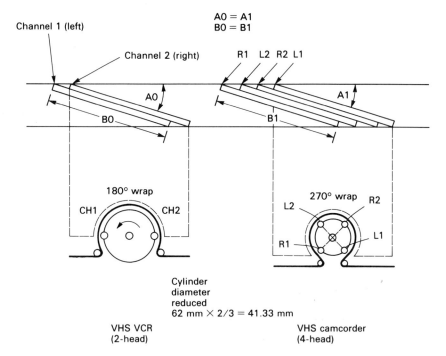

FIGURE 1-4 Comparison of tape-wrap configurations to produce interchangeability between VHS VCR and VHS camcorder tape-record patterns

camcorder, four heads are used, and the tape is wrapped 270° around a 41.3-mm cylinder that rotates at 2700 rpm.

The smaller cylinder is used to make the VCR portion of the camcorder more compact. The 4-head configuration and 270° wrap are used to make the camcorder tape recording pattern compatible with that of standard VHS format recording pattern. Since the camcorder cylinder is exactly two-thirds that of a standard VHS cylinder, three-fourths (or 270°) of the cylinder wrap is required to keep the same track length.

Figure 1-4 shows how the 4-head configuration and 270° tape wrap produce a record pattern that is equal to that of a standard 2-head VHS pattern. As shown, the four video heads record (and play back) in sequence. The typical sequence is L1 (left-head number 1), R2, L2, and R1. This produces the same pattern as the standard VHS left-right pattern.

In the conventional VHS format, where the heads are 180° apart, one head is leaving the tape just as the next head touches the tape. This produces some overlap (about 1.3 ms), which is eliminated during playback by circuits following the head-switching functions.

In the 4-head camcorder configuration, at least three of the heads touch the tape at the same time. As a result, only one head must be turned on during both record and playback. This head-selection process is done by the head-switching circuits.

1-2.1 Head-switching During Record

Figure 1-5 shows typical or generalized head-switching circuits found in VHS camcorders during record. Selection among the four heads is performed by the SW1, SW2, SW3, and SW4 switching signals. These signals actuate gate circuits that control video signals from the camera to the recording heads.

The switching signals hold the gate open for 17.3 ms (or longer) and permit camera signals to be applied at the corresponding head for this period of time. Note that there is some overlap (about 1.3 ms in Fig. 1-5) between L1 and R2, R2 and L2, L2 and R1, and R1 and L1. This is the same overlap as occurs on a conventional VHS VCR between the left and right head-switching.

1-2.2 Head Switching During Playback

Figure 1-6 shows typical or generalized head-switching circuits found in VHS camcorders during playback. Again, at least three of the heads are picking up recorded information from the tape simultaneously, and all four heads are controlled by the head-switching circuits.

For example, during period 1, only SW1 is high, so Q1 turns off and Q2 through Q4 turn on. Under these conditions, the signal picked up by

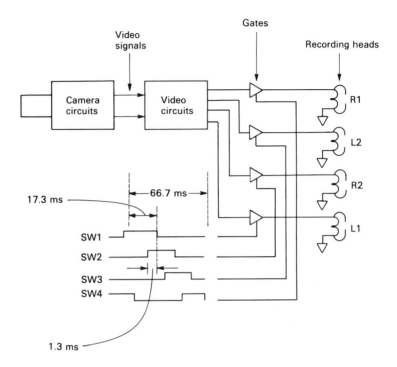

FIGURE 1-5 Typical head-switching circuits in VHS camcorders during record

head L1 is applied through amplifier 1 to the head-switching circuit. The signals at R1, L2, and R2 are shorted to ground through Q2–Q4.

During period 2, only SW2 is high, so only the R2 head output is passed; during period 3, only the L2 output is passed; and during period 4, only the R1 output is passed.

The amplified head-output signals are applied to the playback circuits through the head-select and head-switching circuits. The head-switching circuits are controlled by 16.7-ms pulses. This eliminates any overlap between head signals. The head-select circuits are controlled by pulses of twice that duration, so the signals of only one head pass at a time. Head switching is discussed further in Chapter 6.

1-2.3 Camcorder Tape Path and Drive

Figure 1-7 shows a comparison of the tape path and drive for a VHS camcorder and a standard VHS VCR.

Note that the 270° tape wrap requires a somewhat different path (and much more twisting) than does that for a standard VHS VCR. Also note that for most VHS camcorders, the tape supply reel and takeup reel are

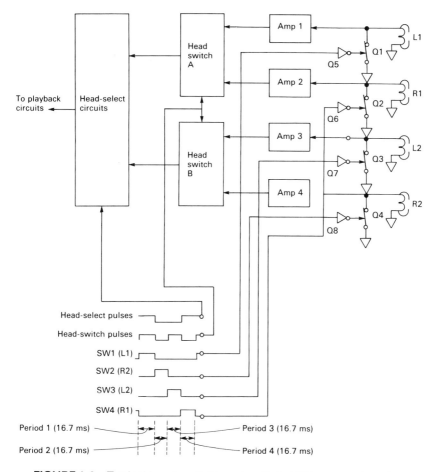

FIGURE 1-6 Typical head-switching circuits in VHS camcorders during playback

rotated by the capstan motor through an idler. A friction clutch reduces the speed so that the reels are rotated at the correct speed in relation to the tape capstan.

1-3 COLOR CAMERA BASICS

The color video cameras used in camcorders are one of the most complex pieces of electronic equipment in the consumer electronics field. To service a color video camera of any kind efficiently, you must understand both optical technology and the operation of the color-processing circuits. For this reason, all of Chapter 4 is devoted to color video cameras. For now, let us review

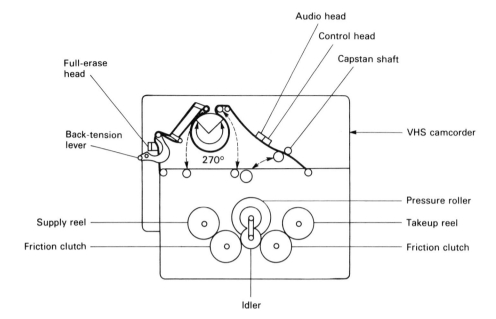

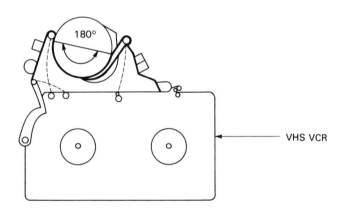

FIGURE 1-7 Comparison of the tape path and drive for a VHS camcorder and a standard VHS VCR.

the basics, such as the principles of light, forming images with electromagnetic deflection, and processing of colors.

1-3.1 The Basic Principles of Color and Light

Figure 1-8 shows the basic principles of color and light.
As shown, color may be produced by one of two processes, the *sub-*

Sec. 1-3   Color Camera Basics   11

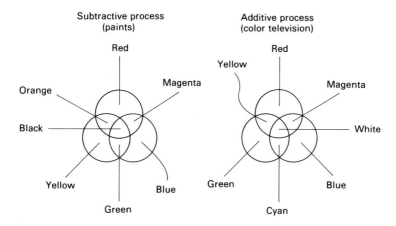

FIGURE 1-8 Basic principles of color and light

tractive process or the *additive process*. The subtractive process is used when working with paints. In the case of light, and color video cameras, the additive process is used exclusively.

The three *primary spectra of light* (or *primary colors* in the additive process) are *red, blue*, and *green*. Using the additive process (color cameras), any desired color can be obtained by *varying the intensity* of the three spectra or primary colors.

In video cameras, the most important factor in the reproduction of color is the type of *light source*. Even if the same object is photographed (or "shot") by a color camera with different light sources, the reproduced color varies according to the light source. That is why the color of an object appears quite different when photographed under incandescent and fluorescent lamps or when photographed on sunny or rainy days.

Light sources are classified as to *color temperature*. Figure 1-9 shows the color temperatures of some common light sources. The color temperature of light is measured in *degrees Kelvin* and refers to the color of light given off by carbon at different temperatures.

To create a natural color in the color camera, regardless of the light source, camcorders (and most color video cameras) are equipped with a *white-balance circuit*. Most camcorders have both an automatic white-balance circuit (for normal color balance) and a manual override (for special color effects).

1-3.2 The Color Pickup

Most camcorders use some form of *color pickup tube* (Saticon, Newvicon, etc.). However, some later-model camcorders use a solid-state MOS (metal-oxide semiconductor) pickup or a CCD (Charge Coupled Device)

Natural Light (°K)	Color Temperature		Artificial Light (°K)	Colors of Light
Fair, blue sky line	—10,000	10,000—	Color television	Bluish
Slightly cloudy sky light	—8,000	8,000—		
Cloudy, rainy sky light	—7,000	7,000—		
	—6,500	6,500—	Fluorescent lamps	
Sunlight in fine-weather, midday	—6,000	6,000—	Daylight camera flashbulb	
	—5,500	5,500—		
Average sunlight in fine weather	—5,000	5,000—	Blue lamp for photography	
Sunlight 2 hr after sunrise and before sunset	—4,500	4,500—	Fluorescent lamp (white)	Whitish
	—4,000	4,000—	Normal camera flashbulb	
Sunlight 1 hr after sunrise and before sunset				
	—3,500	3,500—	Fluorescent lamp (off-white)	
Sunlight 40 min after sunrise and before sunset	—3,200	3,200—	Tungsten lamp (for photography)	
			Halogen lamp	
	—3,000	3,000—	Iodine lamp	Yellowish
	—2,800	2,800—	Normal tungsten lamp	
Sunrise Sunset				
	—2,500	2,500—		
Sunlight 30 min after sunrise and before sunset			Acetylene lamp Kerosene lamp	
Sunlight 20 min after sunrise and before sunset	—2,000	2,000—	Candlelight	Reddish

FIGURE 1-9 Color temperatures of some common light sources

pickup. In any event, the color pickup is a major component in any color camera or camcorder.

Figure 1-10 shows the relationship of the object to be photographed and the pickup. As in the case of any camera, the visual image shot by the camera is presented to the pickup (and ultimately to the viewer) through an

Sec. 1-3   Color Camera Basics

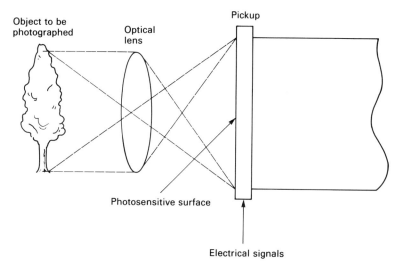

FIGURE 1-10 Relationship of the object to be photographed and the pickup

optical lens. The lens focuses the image on the photosensitive surface of the pickup, which, in turn, converts the image to electrical signals.

1-3.3 The Basic Black and White (B&W) Camera

Figure 1-11 shows the deflection system for a typical video camera pickup tube (the Newvicon). Note that this deflection system is quite similar to that of a TV picture tube. That is, electrostatic deflection is used for focusing the beam, while electromagnetic deflection is used to form the screen raster. This similarity results from the fact that a video camera pickup tube is essentially the reverse of a TV picture tube.

With a TV picture tube, an electron beam strikes light-sensitive material on the inside of the tube surface. The amount of light produced where the beam strikes the tube surface (the *target point*) depends on the intensity of the beam. In turn, the beam intensity is determined by the video signal.

In a video camera, the amount of light at the target point determines the intensity of the signal produced by the camera pickup.

With both TV picture tubes and video camera pickups, the beam is deflected to produce a raster (typically with the EIA standard of 525 lines, 60 fields, and 30 frames) on the tube surface. As a result, the amount of light at any given point on the camera pickup surface produces a corresponding amount of light at the same point on the TV picture tube surface.

In the Newvicon tube of Fig. 1-11, the electron beam from the tube gun is accelerated by grid G2 and then passes through the beam-limiting aperture to generate fine-diameter beams. These beams are then focused by the

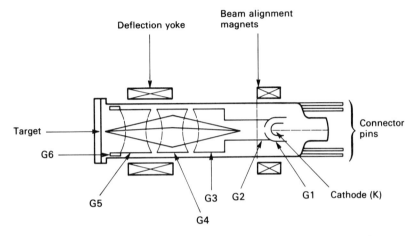

FIGURE 1-11 Deflection system for a typical video camera pickup tube (the Newvicon)

electrostatic lens composed of G3, G4, and G5. Grids G5 and G6 form a collimating lens through which the beams are deflected so that the beams always hit the target at right angles.

Figure 1-12 shows how the video signal is formed and varied by light in a Newvicon tube. The electron beam scans across the target area which is coated with photoconductive material. This produces a raster on the materials (which form an inner layer on the tube surface).

The photoconductive layer creates a number of elements (or the electrical equivalent of those elements). In effect, electrostatic "capacitors" in parallel with LDRs (light-dependent resistors) are formed by the layer materials. All these elements are connected and produce a video signal output when subjected to variations in light.

When there is no light striking the face of the pickup tube, the LDRs create a high resistance. Whenever light hits the face of the target area, the resistance drops at that point. The level of the drop depends on the intensity of the light.

When the beam first scans the target area, each "capacitor" is charged through the circuit loop formed by the beam, load resistance (RL), power source, and photoconductive materials. When the beam is not in contact

Sec. 1-3   Color Camera Basics   15

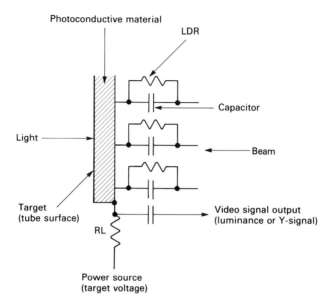

FIGURE 1-12 How the video signal is formed and varied by light in a Newvicon tube

with the element, the "capacitor" slowly discharges through the LDR connected across the capacitor.

Since the LDR resistance varies with changes in light, the capacitor recharges back to the target potential on each scan of the beam. This produces a corresponding charging current at each point. It is the detection of this charging current that produces the video signal. As a result, the video signal intensity corresponds to the light intensity at any given point on the raster. As in the case of a TV picture tube, the black and white video signal is referred to as the *luminance signal* or Y-signal.

1-3.4 The Basic Color Camera

In addition to a luminance signal, a color video camera must also generate a *chrominance* or *chroma* signal to represent the color at any given point on the raster. In the simplest of terms, this is done by separating the colors from each other at the pickup, forming signals for each separated primary color, and then recombining the color signals with the luminance (B&W) signal to produce a color video signal (that is equivalent to the familiar NTSC broadcast signal).

Separation of the colors is done by means of a *stripe filter* located between the incoming light and the target surface. Figure 1-13 shows the composition of a stripe filter as well as how the signals are separated.

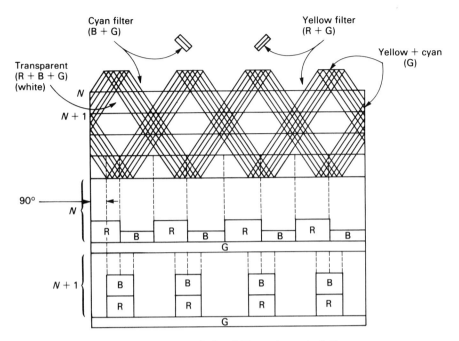

FIGURE 1-13 Striped filter characteristics

Note that the filter, composed of *yellow* and *cyan* stripes, separates the R- (red), B- (blue), and Y- (luminance) signals from each other in the signal output of the pickup tube. The following is a brief description of how this separation is done. Refer to Chapter 4 for a more detailed discussion of color video camera functions.

To see how the colors are separated, consider the signals produced by the pickup when reading out two consecutive lines, identified as N and $N + 1$ in Fig. 1-13.

When the beam crosses the transparent, diamond-shaped space, a white signal is developed (since white is considered equal parts of red, green, and blue). In effect, all colors pass and produce corresponding signals.

When the beam crosses the yellow stripe, a red-plus-green signal is produced, but blue is not passed. When the beam crosses the cyan stripe, a blue-plus-green signal is produced, but red is not passed.

When the beam crosses an area where the yellow and cyan stripes cross, the yellow stripe blocks blue, while the cyan stripe blocks red. As a result, only green passes.

Note that the two consecutive lines, N and $N + 1$, have a phase difference of 90°. This is important to remember since the next step in producing a color signal is to combine the signals from all lines in consecutive order.

As shown in Fig. 1-14, signals from the pickup are applied through a preamp to a low-pass filter (LPF) and a bandpass filter (BPF). The LPF

Sec. 1-3   Color Camera Basics   17

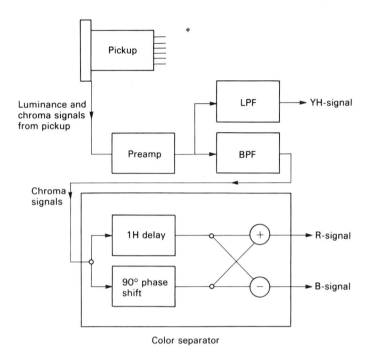

FIGURE 1-14 Basic color separation process

separates the luminance signals, identified as YH (or Y-high, high-frequency luminance) at this point, while the BPF passes the color signals. (If you are not already aware, different colors produce signals of different frequencies.)

The color signals are applied to a 90° phase-shift network and a 1H delay network. (It is assumed that you are familiar with the term 1H, which refers to the period of time required to produce a horizontal line on a TV screen, or approximately 63.5 μs.)

The red or R signal is obtained by adding the 1H and 90° signals. The blue or B-signal is obtained by subtracting the 1H and 90° signals. The red, blue, and luminance signals are then combined to produce the equivalent of an NTSC signal.

1-3.5 Producing an NTSC Video Signal from the Basic Camera Signals

Figure 1-15 shows how the basic color camera signals (produced by the pickup, either tube or solid state) are combined to produce the video signal (that is recorded on tape by the VCR portion of the camcorder).

In this system, the brightness of the video signal (at any given point on the raster) is determined by the level of the luminance signal. The color information is converted into a *color difference signal* (R-YL and B-YL).

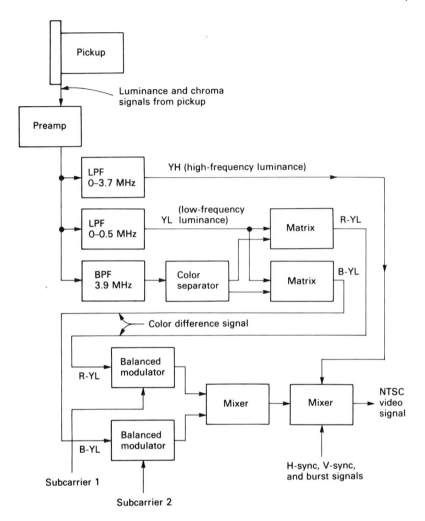

FIGURE 1-15 How the basic color camera signals are combined to produce the video signal

Note that the luminance signal is split into two signals, YH and YL. YH (Y-high) is in the frequency range from zero to 3.7 MHz, while YL (Y-low) is in the range from zero to 0.5 MHz. The YL-signal is combined with the R and B color signals to produce R-YL- and B-YL-signals. In turn, the R-YL- and B-YL-signals are combined with the YH-signal to produce the composite NTSC signal.

As shown in Fig. 1-15, after the R- and B-signals are passed through the BPF and color separator (Fig. 1-14), the R- and B-signals are combined

Sec. 1-4 Typical Camcorder Features   19

with YL in a matrix. The outputs from the matrix are the color difference signals R-YL and B-YL.

The R-YL- and B-YL-signals are amplitude modulated by two 3.58-MHz subcarriers. (The color subcarriers are similar to the 3.58-MHz color burst signal in TV and are 90° apart.)

The modulated R-YL- and B-YL-signals are then combined in a mixer. This process is known as *quadrature modulation*. The output of the mixer is further combined with the YH-signal in another mixer. Both horizontal and vertical sync signals are added in this second mixer to produce the composite NTSC signal.

1-4 TYPICAL CAMCORDER FEATURES

Figure 1-16 shows features found on a typical VHS camcorder. It is essential that you be aware of these features to understand properly the circuits used to provide the features. Keep in mind that not all VHS camcorders have all the features described here.

1-4.1 General Information

As discussed, a camcorder combines the functions of a video camera and a VCR in one lightweight, easy-to-use instrument. This eliminates interconnecting cables and simplifies operation.

A typical camcorder can record directly from the built-in video camera or, with an optional *audio/video input adapter*, from an audio/video source such as another VHS VCR. The audio/video played back by the camcorder, or fed directly from the camera portion of the camcorder, may be viewed and heard on a TV receiver/monitor by feeding the camcorder audio/video to the TV inputs through an optional *audio/video output connector cable* or by using an optional RF adapter.

Because a VHS camcorder uses standard VHS video cassettes, camcorder tapes can be played back on the camcorder or any other VHS VCR.

In addition to being easy to use, lightweight, and compatible with other VHS recorders, a typical VHS camcorder is equipped with autofocus, power zoom, automatic white balance, electronic viewfinder, built-in microphone, and quick review. Let us go through some of these features in more detail.

1-4.2 Pickup

Most VHS camcorders have a $\frac{1}{2}$-in. pickup tube (Saticon, Newvicon, etc.), although some of the later-model instruments use a $\frac{2}{3}$-in. solid-state MOS pickup. With either pickup, the camcorder can be operated indoors

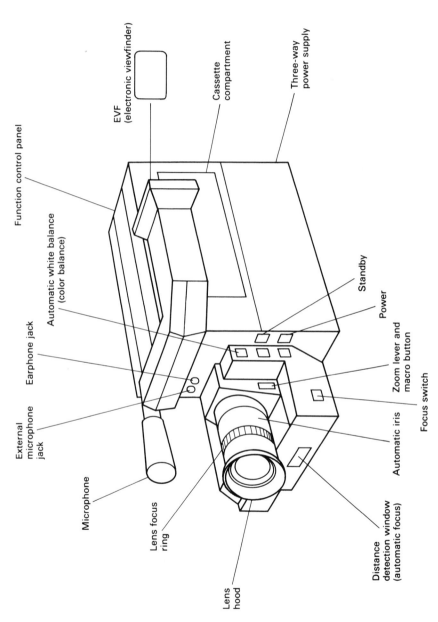

FIGURE 1-16 Features found on a typical VHS camcorder

Sec. 1-4 Typical Camcorder Features   21

or out with light levels as low as about 7 to 10 lux. The optimum light level for a typical VHS camcorder is about 1500 lux (10 lux = 1 footcandle).

1-4.3 Electronic Viewfinder

An electronic viewfinder (EVF) displays (in black and white) exactly what is recorded on the tape (in color). The EVF also doubles as a B&W monitor for viewing "instant replays" after taping. On most EVFs, an LED signals when the camcorder is in the record mode and when battery level is low.

Some EVFs have an *adjustable diopter* that allows EVF focus to be changed without affecting lens focus. This permits users who wear eyeglasses to move the eyeglasses when operating the camcorder.

Many EVFs include a number of displays or graphic indicators in addition to record status and battery condition. Typical display indicators include indoor/outdoor light setting, time and date, tape counter, memory-on, low-light warning, tape warning, and dew warning.

1-4.4 Power Supply

A typical camcorder powers itself from three different sources. This makes it possible to move video production virtually anywhere.

Indoors, the a-c adapter/charger allows you to plug in the camcorder, charge the batteries, or hook up to a monitor TV or standard TV. Most adapter/chargers operate on 110 or 240 V, 50 or 60 Hz, permitting you to operate anywhere in the world. (However, you may need a plug adapter in some countries.)

Outdoors, with a single battery pack snapped to the camcorder, you have about 2 hr of continuous use. Often, there is a *standby mode* to minimize power consumption when operating on batteries.

In a vehicle, the 12-V auto electrical system can power the camcorder through a d-c car cord (generally an optional accessory).

1-4.5 Lens

A typical VHS camcorder has an f1.2 lens with a 6-to-1 *zoom ratio*. The zoom can be manual or motorized. A hand-grip control lets you zoom in for close-ups or zoom out for panoramic shots. The lens stops at the desired perspective when the hand control is released. The zoom ratios can also be adjusted manually by rotating the lens ring.

Most camcorder lenses also include a *macro* function that permits you to obtain sharp images close up (from about $\frac{3}{8}$ in. to 1 in.). This gives a sharp, enlarged image for shooting small objects without loss of detail.

1-4.6 Automatic White Balance

As discussed in Sec. 1-3.1, white balance is critical to obtaining proper colors. Most camcorders have circuits that continuously adjust for proper white balance (also called *color balance* in some camcorder literature). A manual white-balance (or color-balance) control is included for creating special visual effects or for unusual lighting conditions.

1-4.7 Automatic Focus

An automatic focus system, using an infrared light beam aimed at objects positioned before the center of the lens, keeps moving objects in focus. This automatically maintains a sharp image, even during zooms. The automatic focus can be switched off for manual focus control (generally to create special effects).

1-4.8 Automatic Iris

To assure correct exposure (for proper picture brightness and contrast), the camcorder automatically responds to available light conditions and adjusts the aperture accordingly. A manual override is provided for unusual lighting conditions.

1-4.9 Microphone and Earphone

Most camcorder microphones are front-mounted for increased sensitivity to audio from the camcorder subject. Some camcorders also include an accessory jack for an optional microphone. Usually, camcorders also include an earphone jack to monitor both record and playback.

1-4.10 High-quality VHS (HQ)

Many late-model camcorders have circuits to enhance the VHS picture quality by sharpening image definitions. This allows complete compatibility with recordings made on other VHS machines.

1-4.11 Pause and Quick Review

Many camcorders have pause and quick review functions. In the RECORD mode, the pause function is controlled by a pushbutton. You press the pushbutton to start recording a scene and press again to stop. For a quick look at the last 3 to 4 sec of the previously recorded scene, you press the review button (with the camcorder in the PAUSE mode).

1-4.12 Search and Stop Action

On camcorders with search and stop function, you use the EVF to scan forward or reverse through recorded material to find desired program segments. Or you can examine details easily missed during recording by stopping the action at any point on the tape. These special effects can also be used when viewing tapes on a monitor or standard TV.

1-4.13 Input/output Connections

Most camcorders feature complete flexibility for use with external components. (We describe some typical examples in Chapter 2.)

Audio/video output connections send playback signals to a VCR or monitor TV for dubbing or viewing, using cables. Likewise, the camcorder output can be applied to a conventional TV through an RF output adapter.

Audio/video input connections permit recording (in the SP mode only) from selected external sources, such as a VCR, tuner, and so on, using an optional adapter.

1-5 TYPICAL CAMCORDER SPECIFICATIONS

The following specifications for a "typical" VHS camcorder are included here for reference.

Power source	12-V d-c, or with an adapter, 110/120/220/240-V a-c, 50/60 Hz
Power consumption	12-W d-c, 37-W a-c (adapter)
Television system	EIA standard (525 lines, 60 fields, 30 frames) NTSC color signal
Video recording system	4 rotary heads, helical scanning system Luminance: FM azimuth recording Color signal: converted subcarrier phase-shift recording
Audio track	1 track
Tape format	Tape width $\frac{1}{2}$-in. (12.7 mm), high-density tape
Tape speed	$1\frac{5}{16}$ ips (33.35 mm/s) in SP mode
Record/playback time	160 min with T160 used in SP mode
FF/REW time	Less than 10 min with T120
Camera tube	One $\frac{1}{2}$-in., integrated stripe filter, electrostatic focus, magnetic deflection, separate mesh, special tube Newvicon, type S4161

Focus	Electrostatic
Lens	6-to-1 zoom lens, automatic iris (closed with power off), autozoom lens and macro construction, autofocus system, f1.2 (f: 9 mm–54 mm), lens filter diameter 49 mm
Electronic viewfinder	Monochrome $\frac{1}{2}$-in. CRT (built-in)
Minimum light intensity on optical image	10 lux (at f1.2), 1 footcandle
Optimum light intensity on optical image	1500 lux, 150 footcandles
External microphone input level	−70 dB, 600-ohm unbalanced, M3 connector
Operating temperature	32°F–104°F (0°C–40°C)
Operating humidity	10–75%
Weight	7.7 lb (3.5 kg) (with internal battery pack)
Dimensions (in.)	$6\frac{1}{8}$ W × $7\frac{1}{2}$ H × $13\frac{3}{4}$ D

2

USER CONTROLS, OPERATION PROCEDURES, ACCESSORIES, and INTERCONNECTIONS

This chapter describes the basic user controls and operating procedures for typical VHS camcorders. We concentrate on the functions of the controls, to establish a basis for troubleshooting, rather than on camera techniques. A careful study of the procedures will help you to understand operation of the camcorder circuits. We also describe typical camcorder accessories and interconnection procedures.

Keep in mind that you must study the operating controls and indicators for any camcorder you are troubleshooting. This book describes "typical" controls and indicators, but there are subtle differences in operation you must consider.

For example, most camcorders have displays that show a low-battery condition and a tape count. In some camcorders, these displays are in the form of LED or LCD readouts, while in other camcorders both displays appear on the viewfinder screen. Further, the viewfinder screen of some (but not all) camcorders shows the recording time remaining and date/time. Keep in mind that there is nothing more frustrating than troubleshooting a failure symptom when the camcorder is supposed to work that way!

2-1 ACCESSORIES

Figure 2-1 shows the accessories for a typical VHS camcorder. Note that some of these accessories are optional, while others are usually available with all camcorders.

In normal outdoor use, the camcorder is powered by a battery pack clipped onto the camcorder as shown. Typically, the battery is a 12-V Ni-

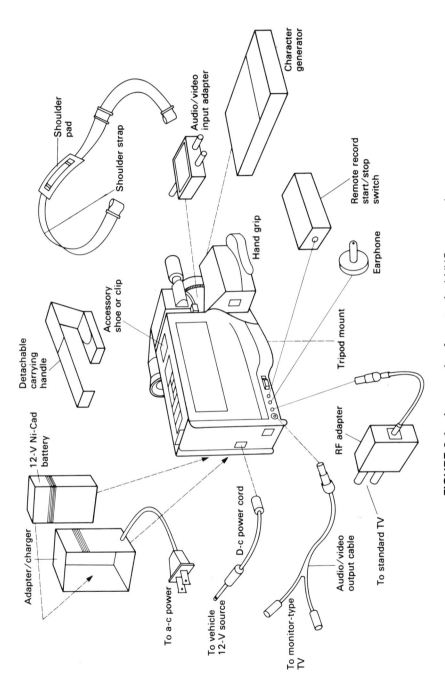

FIGURE 2-1 Accessories for a typical VHS camcorder

Cad type that can be recharged by an a-c *adapter/charger* unit. Some camcorders have a *standard battery* and an optional *long-play battery*. With some camcorders, it is also possible to charge the battery from a *battery-charging cord* instead of the adapter/charger.

For indoor use, the camcorder can be powered from an a-c source (wall outlet) through the adapter/charger. Note that with most camcorders, it is not possible to charge the battery and power the camcorder simultaneously.

When used in a vehicle, the camcorder can be powered by the vehicle's 12-V source through a d-c *power cord* connected to the camcorder in place of a battery or adapter/charger.

When used with a monitor-type TV, the audio/video being recorded or played back by the camcorder is displayed on the monitor through the *audio/video output cable* or adapter. When used with a standard TV, the audio and video are converted to an unused channel (typically channel 3 or 4) by an *RF adapter*. (This is essentially the same as the RF unit or modulator in a VCR.)

On some camcorders, the RF adapter is part of the adapter/charger. Also, some camcorders include a 300/75-ohm matching transformer for TV sets with 300-ohm antenna terminals instead of coax inputs.

Most camcorders have an *audio/video input adapter* or cable that makes it possible to record from another audio/video source (such as another VHS VCR).

Most camcorders have an *accessory earphone* to monitor the audio being recorded or played back). Likewise, some camcorders have a microphone input jack for an *auxiliary microphone* (in addition to the built-in microphone).

Many camcorders include a *remote record start/stop switch*. This makes it possible to start and stop recording from a remote location.

Some camcorders have an optional *character generator*. This makes it possible to add titles to the video.

There is often a *tripod mount* on the bottom of the camcorder and an *accessory shoe* or *clip* at the top. The shoe can be used for a carrying handle or to mount other accessories. Generally, most camcorders have a *shoulder strap* and possibly a *shoulder pad*.

In addition to the standard accessories described thus far, many late-model camcorders come with such optional accessories as lens filters and caps, lens extenders, wide-angle adapters, film/slide adapters, carrying case (both hard and soft), and wireless microphone.

2-2 TYPICAL OPERATING CONTROLS AND INDICATORS

Figures 2-2 through 2-5 show the operating controls and indicators for two typical VHS camcorders. The following paragraphs describe the control and indicator functions.

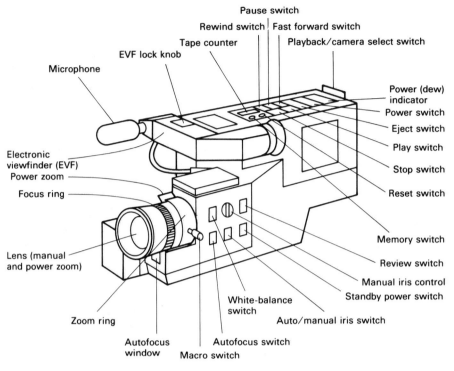
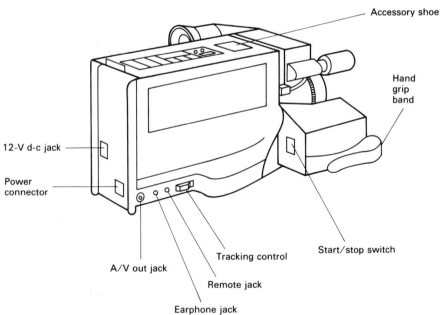

FIGURE 2-2 Operating controls and indicators for typical VHS camcorder

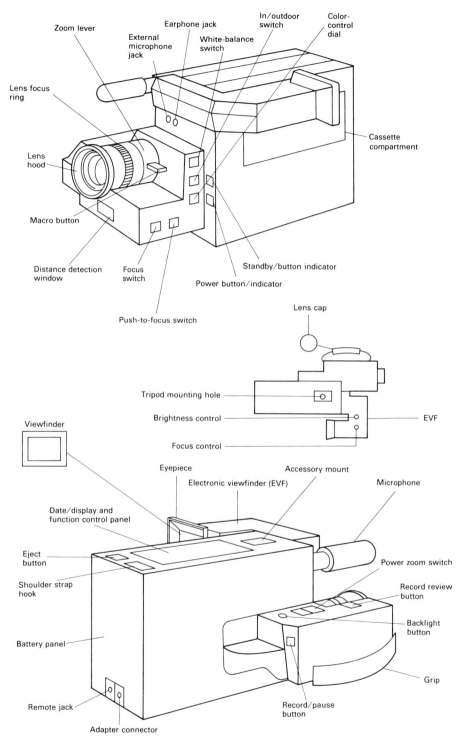

FIGURE 2-3 Alternate operating controls and indicators for typical VHS camcorder

30 User Controls, Operation Procedures, Accessories, and Interconnections Chap. 2

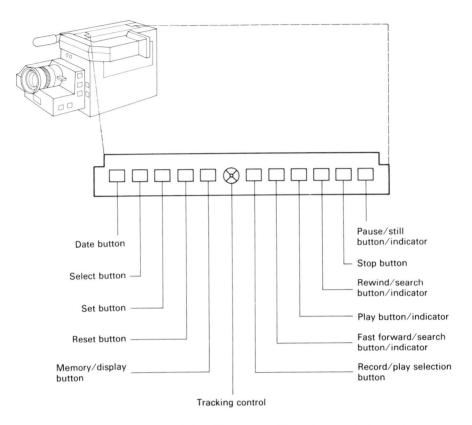

FIGURE 2-4 Details of date/display and function control panel

2-2.1 Controls and Indicators for the Camcorder of Fig. 2-2

The *12-Vd-c jack* may be used to operate the camcorder from a car battery by plugging an optional car cord into the cigarette lighter.

The *A/V out jack* receives the accessory audio/video connector or cord, which allows video and audio out to another device (typically a monitor TV). The audio/video output can also be connected to a standard TV through an RF adapter.

The *accessory shoe* is used to attach an optional wireless microphone or other lightweight video accessory.

The *autofocus switch* allows selection of manual or autofocusing.

The *power indicator* turns on to indicate that power has been applied. The power indicator also functions as a *dew indicator* (by flashing to indicate excessive moisture in the recorder portion of the camcorder). When the

Sec. 2-2 Typical Operating Controls and Indicators   31

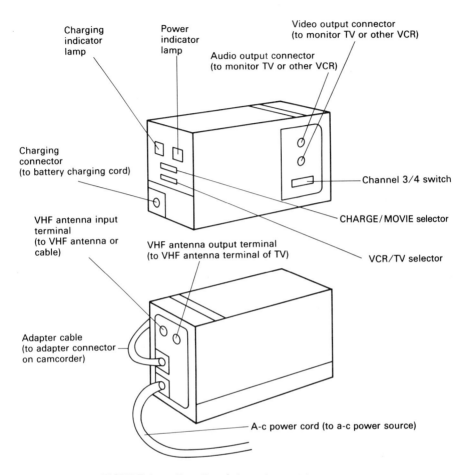

FIGURE 2-5 Details of the adapter/charger

power indicator flashes, the camcorder will not operate (as is the case with most VCRs).

The *EVF lock knob* allows the electronic viewfinder to be positioned for comfort and viewing ease. The electronic viewfinders of most camcorders are spring loaded and can be moved from a stored or at-rest position to the normal viewing position.

The *earphone jack* is used to connect an earphone to the camcorder.

The *eject switch* is pressed to remove or insert a cassette tape.

The *electronic viewfinder* (EVF) displays the scene being observed by the camera.

The *autofocus window* receives an infrared signal that controls the autofocus system. The infrared signal (transmitted through the camera lens) is reflected from the object located in front of the camera and is returned to the autofocus system through the autofocus window. As discussed in Sec. 2-2.2, the autofocus window is also called the *distance detection window* in some camcorders.

Press the *fast forward* switch while in the STOP mode to advance the tape rapidly in the forward direction. Press the fast forward switch while in the PLAYBACK mode to scan the recorded material visibly in the forward direction. (This is also called the *search function* in some camcorders.)

With the *autofocus switch* in the MANUAL position, the camera can be focused by viewing the picture displayed on the EVF while adjusting the *focus ring* for proper focus.

With the auto/manual iris switch in AUTO, the camera automatically adjusts the iris. Set the switch to MANUAL when the iris is to be adjusted manually, with the *manual iris control*. (Note that not all camcorders have a manual iris control function. Iris control is always automatic on such camcorders.)

The adjustable *hand grip band* is provided for comfort and stability during camera use.

The f1.2, 6-to-1 *power zoom lens* directs the incoming light onto the camera tube. The picture size can be magnified six times with the zoom feature.

For manual zoom, set the autofocus switch to MANUAL, and rotate the *zoom ring* in one direction for close-up (T, telephoto) or in the other direction for wide-angle (W) pictures. For *macro close-ups*, set the autofocus switch to MANUAL, and rotate the zoom ring in the wide-angle direction, while pulling the *macro switch* located on the *zoom lever*.

The zoom feature is also motor driven and controlled by the telephone/wide-angle switches. When the telephoto (T) switch is pressed, the zoom ring moves in the telephoto direction, providing a close-up view of the subject. When the wide-angle (W) switch is pressed, the zoom ring moves in the wide-angle direction, increasing the area of the scene.

A digital *tape counter* indicates the relative position of the program on the tape. Press the *reset* button to reset the counter display to 0000. When the *memory* switch is on during rewind, the tape moves to an indication of 0000 on the counter.

The *microphone* is located on the front of the camcorder and is most sensitive to sounds coming from the direction in which the lens is pointed.

Press the *power switch* to apply or remove power. The *power connector* receives the adapter/charger or the battery, to power the camcorder. When the *standby power* switch is set to STANDBY, power consumption is minimized. The camera operates normally when the standby power switch is set to ON.

Press the *play switch* to play back recorded material. Press the *stop switch* to end playback operation. Press the *pause switch* for momentary pause during playback operation. The *playback/camera select switch* must

Sec. 2-2   Typical Operating Controls and Indicators   33

be in PLAYBACK to play back a previously recorded tape and in CAMERA for camera operation.

The *start/stop switch* is used to toggle between PAUSE and RECORD. Press start/stop once, and the camcorder switches from pause to record (causing the record indicator in the viewfinder to turn on). Press start/stop again, and the EVF record indicator turns off (indicating that the camcorder is in pause mode).

When the *rewind switch* is pressed in the stop mode, the tape is rewound. When the rewind switch is pressed in playback, it is possible to scan the recorded material visibly in the reverse direction. (This is also called the *reverse search* function in some camcorders.)

Press the *review button* during camera pause to view the last 4 sec of a recorded segment.

Proper color balance is maintained automatically when the *white-balance* switch is set to AUTO. For special lighting conditions, the white balance may be set manually by holding the white-balance switch in AUTO SET until the record indicator in the EVF stops flashing. The white balance then remains in the selected condition until the white balance is set to AUTO. Note that on some camcorders, the white-balance switch can be set to a manual position and varied by a control to get a desired color condition.

The *tracking control* provides the familiar tracking function found on most VCRs. When playing prerecorded tapes, or tapes recorded on other units, black and/or white streaks may appear on the TV screen. The tracking control is adjusted slowly in either direction until the streaks disappear.

The *remote jack* is used to connect the accessory remote start/stop switch.

2-2.2   Controls and Indicators for the Camcorder of Figs. 2-3 through 2-5

The *distance detection window* detects subject-to-camera distance for the automatic focus feature. As discussed in Sec. 2-2.1, the distance detection window is also called the autofocus window in some camcorders.

The *lens hood* shields the camera lens and accepts the *lens cap* to protect the lens.

The *lens focus ring* is rotated to focus the picture manually when the *focus switch* is in the MANUAL position.

The *external microphone jack* permits use of an external microphone for audio pickup closer to the subject. The *built-in microphone* is automatically disconnected when an external microphone (600-ohm impedance) is plugged in.

An accessory earphone can be plugged into the *earphone jack* to monitor sound while recording or while playing back recordings.

The *in/outdoor switch* is set for indoor or outdoor lighting conditions as required. Then, the *white-balance switch* is set at the AUTO position to maintain optimum color balance automatically in either indoor or outdoor settings.

When either red or blue dominate the lighting conditions, set the *white-balance switch* to FIXED, and rotate the *color-control dial* until the colors are best (as indicated by a monitor TV connected to the camcorder). The color-control dial is normally left in the center or detent position. The dial must be pushed in to rotate.

When the *standby button/indicator* is pressed, the camcorder is ready to record, but uses a minimum amount of battery power.

The *power button/indicator* turns power on and off.

The *macro button* (on the *zoom lever*) is used to unlock the zoom lever so that the lever can be moved into the far left macro position for close-up photography. Typically, the subject can be as close as 1 in. from the lens (when set to the macro position).

The *zoom lever* is used for manual zoom operation. The *power zoom switch* operates the zoom drive motor to zoom in (telephoto or T) or zoom out (wide-angle or W).

When the *focus switch* is in AUTO, the lens is focused automatically. When the focus switch is in MANUAL, the lens is focused manually with the lens focus ring. When the focus switch is in MANUAL, and the push-to-focus switch is pressed, the lens is focused automatically (until the push-to-focus switch is released).

The *remote jack* provides for connection of a wired remote control.

The *accessory mount* provides a convenient mounting facility for the handle and optional lights.

Press the *record review button* to play back the last few seconds of a recording.

Press the *record/pause button* to start or pause the camera while recording.

When the *backlight button* is pressed during record, the lens opening automatically compensates for high-contrast light conditions.

The *brightness control* and *focus control* adjust the brightness and focus of the viewfinder picture. These are not true operating controls, but are screwdriver adjustments accessible to the user.

Details of the *date/display and function control panel* are shown in Fig. 2-4.

When the *date button* is pressed, the date is displayed (on the EVF) and recorded.

The *select button* shifts the cursor during date selection for day, month, or year.

The *set button* selects numbers while setting in a date.

The *reset button* resets the tape counter (in the EVF) to 0000.

When the *memory/display* button is in the M (memory) position, the tape stops when the tape counter reaches 0000, during fast forward or rewind. The memory/display button also clears the display from the EVF screen.

The *record/play selection button/indicator* selects either the RECORD/PAUSE or PLAY/PAUSE modes. RECORD/PLAY is selected when the indicator is on. Pushing the button again turns the indicator off to indicate that the PLAY mode is selected.

Press the *play button/indicator* to play back recorded material.

Press the *stop button* to end playback operation or release the PAUSE mode (during record).

Press the *pause/still button/indicator* to pause during record or to view a still picture during playback.

Press the *rewind/search button/indicator* to rewind the tape (in the stop mode). If PLAY mode is selected, and the rewind/search button is pressed, the tape is played back (in the reverse direction) on the EVF at about three times normal speed.

Press the *fast forward/search button/indicator* to advance the tape rapidly in the forward direction (in the STOP mode). If PLAY mode is selected, and the fast forward/search button is pressed, the tape is played back (in the forward direction) on the EVF at about three times normal speed.

The *tracking* control is normally left in the detent position, but may be adjusted as necessary to eliminate interference (typically streaks) when playing tapes recorded on other instruments.

Details of the *adapter/charger* are shown in Fig. 2-5. Note that this particular adapter/charger provides for connection of the camcorder to a TV set, in addition to charging the battery.

The *power indicator lamp* turns on when the adapter/charger power cord is connected. The *charging indicator lamp* turns on when a battery is being charged and turns off when the battery is fully charged. The *charging connector* is used when a battery is charged through a *battery-charging cord* (instead of attaching the battery to the adapter/charger).

The VCR/TV and charge/movie switches are interlocked to select either CHARGE and TV (where a battery is being charged and the TV is used for normal viewing) or VCR and MOVIE (where the camcorder is being powered from the adapter/charger and the TV is connected to the camcorder).

The *video output* and *audio output* connectors are for use with a monitor TV.

The *channel 3/4 switch* sets the adapter/charger output to either channel 3 or 4 (whichever is not used for TV broadcasts in the area) and provides the best picture during playback.

The *VHF antenna input terminal* connects the adapter/charger to a VHF antenna or cable. The *VHF antenna output terminal* connects the adapter/charger to the VHF input of a standard TV. The VHF terminals are not used for a monitor TV.

36 User Controls, Operation Procedures, Accessories, and Interconnections   Chap. 2

The *adapter cable* connects the adapter/charger power output to the camcorder (at the adapter connector, Fig. 2-3).

2-3 INTERCONNECTIONS

A camcorder must be properly interconnected for correct operation. Although the following notes apply to specific camcorders (Sylvania, Magnavox, RCA), the connections are typical for many other units.

Typical interconnections include connecting the camcorder to power sources other than the battery, connecting to a TV (both standard and monitor), charging the battery, and connecting to a remote switch and/or character generator.

2-3.1 Connecting to Power Sources Other Than the Battery

Figure 2-6 shows some typical connections to alternate power sources (wall outlet and car battery). Note that on this particular camcorder (Sylvania, Magnavox), the adapter/charger also provides for connection to the

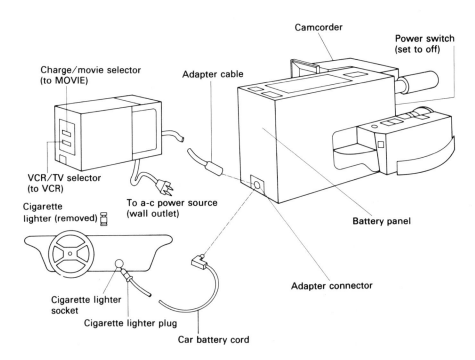

FIGURE 2-6   Typical connections to alternate power sources

TV. Refer to Sec. 2-3.2. In camcorders where the adapter/charger does not provide for connection to the TV (e.g., RCA), the adapter/charger is simply clipped onto the camcorder in place of the battery instead of being connected through an adapter cable.

The following notes apply when using the adapter/charger as shown in Fig. 2-6.

Be sure to turn the power off before making any connections.

Attach the adapter cable of the adapter/charger to the adapter connector of the camcorder.

Connect the power cord to an a-c outlet. Make sure that the power cord plug matches the wall outlet. Use a power plug adapter as necessary. Note that most adapter/chargers operate on line voltages of 110 to 240 V, at frequencies of 50 to 60 Hz, and produce the correct d-c output voltage without a change of switch settings. The voltage/frequency changeover is done automatically by the adapter/charger circuits.

Set the charge/movie selector to MOVIE (and the VCR/TV selector to VCR). The camcorder should now be ready to operate as described in Sec. 2-4.

The following notes apply when using the car battery cord as shown in Fig. 2-6.

Be sure to turn the power off before making any connections.

Insert the car battery cord into the adapter connector at the bottom of the battery panel of the camcorder.

Take the cigarette lighter out of the lighter socket.

Start the car engine, and then insert the cigarette lighter plug of the car battery cord into the socket. Be sure to start the car engine *before* inserting the cigarette lighter plug into the lighter socket. If the engine is not running, the cigarette lighter plug may burn out when the engine is started. If the fuse (in the cigarette lighter plug) does burn out, be sure to use the exact current rating (amperes) for replacement (which is a good idea for most fuse replacements!).

Disconnect the cigarette lighter plug when the camcorder is not being used. This will help you to remember that the engine should be started *before* you insert the plug.

The car engine should be running while the camcorder is in operation. Make sure that the car is in an open, well-ventilated area.

2-3.2 *Connecting to the TV*

Figure 2-7 shows some typical connections between the camcorder and a TV set (both standard and monitor types). Note that on this particular camcorder, the adapter/charger also provides for connection to the TV. Figure 2-8 shows connections made through an RF adapter and audio/video cables.

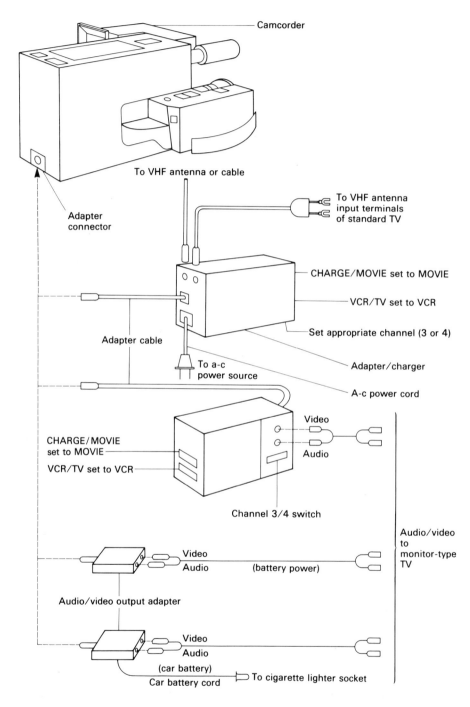

FIGURE 2-7 Typical connections between a camcorder and a TV set

Sec. 2-3   Interconnections   39

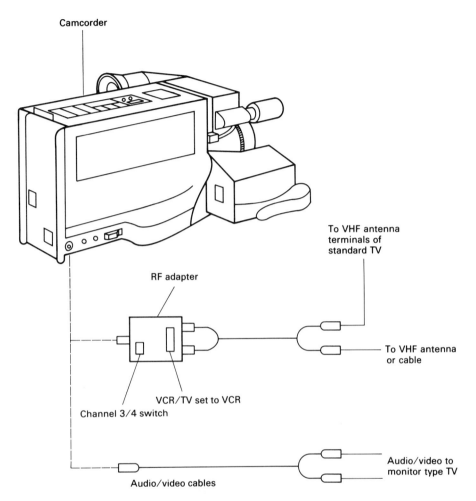

FIGURE 2-8 Typical connections between a camcorder and a TV set (with RF adapter or audio/video cables)

When using the adapter/charger of Fig. 2-7, make sure that the charge/movie switch is set to MOVIE (and that the VCR/TV switch is set to VCR). When using the interconnections of either Fig. 2-7 or 2-8, make certain to select the appropriate channel (3 or 4) on *both* the adapter/charger (or RF adapter) and the TV.

2-3.3 *Charging the Battery*

Figure 2-9 shows some typical connections for charging the battery. Note that in some camcorders, the battery must be clipped to the adapter/charger and cannot be charged through a battery-charging cord as shown.

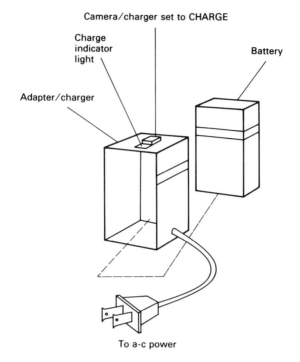
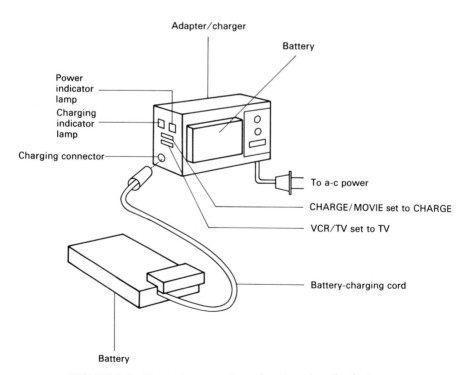

FIGURE 2-9 Typical connections for charging the battery

Sec. 2-3   Interconnections   41

It is possible to charge two battery packs simultaneously with the adapter/charger of Fig. 2-9. One pack is clipped to the adapter/charger, while the other battery pack is charged through the battery charging cord.

The following notes apply when using the adapter/charger as shown in Fig. 2-9.

Connect the battery pack using the battery-charging cord, or place the battery pack in the adapter, or use both methods to charge two battery packs simultaneously.

Connect the power plug of the adapter/charger to an a-c outlet.

Set the charge/movie selector to CHARGE (and the VCR/TV selector to TV). On adapter/chargers where the battery is clipped in place, and a charging cord is not used (such as with RCA), set the camera/charger select switch to CHARGE.

Once the charger is turned on, a charging indicator should also turn on. When charging is complete, the charging indicator should turn off. Battery charging usually takes between 1 and 2 hr. The time increases to between 3 and 4 hr when two battery packs are charged simultaneously.

The following notes on battery charging and batteries apply to virtually all camcorders.

If the charge/movie selector is set to MOVIE (or the camera/charger select switch is set to CAMERA), the battery will not be charged.

The camcorder battery panel cannot be used to charge batteries!

The useful operation time of any battery pack gradually decreases after repeated use and recharging. The battery pack should generally be discarded if the operating time is too short, even after a sufficient charge. However, do not get rid of a battery pack until you have tried a substitute pack (known to be good) under the same charge time/operation time conditions. You may have a problem in the camcorder that is draining the battery beyond normal capacity.

Store battery packs in cool, dry, dark places.

Do not drop or subject battery packs to strong jolts.

Always charge the battery pack soon after use. When batteries are left discharged for a long time, battery life can be shortened (and it may be impossible to recharge the battery). This caution applies mostly to batteries that are fully discharged and left in that condition for a long time.

Do not place the battery pack near heat. This can short-circuit the pack. Never dispose of a battery pack in a fire. (The battery can explode!)

Do not use any charger other than the adapter/charger supplied with the camcorder, and do not use the battery pack for appliances other than the camcorder.

Do not use a battery below 0°C (32°F) or above 40°C (104°F). On most battery packs, a safety device automatically prevents operation of the pack if the temperature exceeds these limits.

Do not use an insufficiently charged or worn-out battery pack.

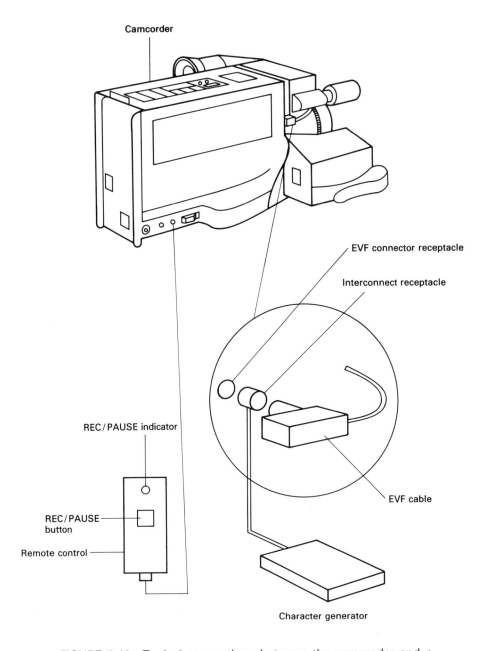

FIGURE 2-10 Typical connections between the camcorder and a remote switch and/or character generator

Sec. 2-4   Typical Operating Procedures   43

Be sure to follow any instructions or battery pack mounting arrows on the camcorder when installing and removing the battery. In many cases, you must operate a *battery release* button to both install and remove the battery.

WARNING: Should the battery pack be broken, and the internal electrolyte-containing sulfuric acid get on the skin or clothing, wash the affected area in water immediately. Then neutralize the acid with sodium bicarbonate (or ammonium water, if available). CALL A DOCTOR OR OTHER LOCAL EMERGENCY MEDICAL AID FACILITY!

2-3.4 Connecting a Remote Switch and/or Character Generator

Figure 2-10 shows some typical connections between the camcorder and a remote switch and/or character generator. Note that most VHS camcorders can be operated with a remote switch, but not all camcorders will accept a character generator.

When a character generator is used, the generator is usually plugged into the EVF connector receptacle on the camcorder in place of the cable from the EVF. In turn, the EVF cable is plugged into an interconnect receptacle on the character generator. Since these connections apply only to certain camcorders, we do not dwell on them in this book.

In the camcorder of Fig. 2-10, the remote control has only one operating control and one indicator. Pushing the rec/pause button starts the recording process, when the camcorder is in the RECORD/PAUSE mode. When pushed again, the rec/pause button resets the camcorder to PAUSE. The rec/pause indicator turns on when the camcorder is in RECORD and goes off in PAUSE.

2-4 TYPICAL OPERATING PROCEDURES

The following notes apply specifically to VHS camcorders with operating controls and indicators similar to those of Figs. 2-3 through 2-5. However, most VHS camcorders have such controls and indicators (possibly with different names), so you can relate these procedures to the camcorder you are servicing.

2-4.1 Typical EVF Displays

Figure 2-11 shows some typical EVF displays. Keep in mind that most EVFs have two functions. First, the EVF is a miniature monitor that displays (in black and white) what is being recorded and what you will see (in color) during playback. The EVF also provides for an "instant replay." Second,

44 User Controls, Operation Procedures, Accessories, and Interconnections   Chap. 2

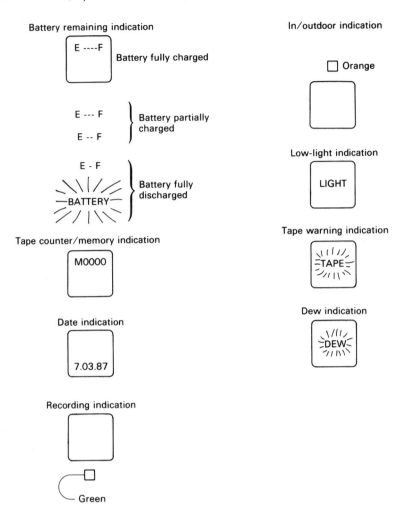

FIGURE 2-11 Typical EVF displays

the EVF display shows camcorder conditions (tape count, battery condition, etc.) without having to remove your eye from the display.

The *battery remaining indication* is not recorded on tape, but shows the condition of battery charge. As illustrated in Fig. 2-11, the number of dashes between the letters E (empty) and F (full) show the battery charge. The number of dashes decrease as the battery is discharged. When the battery is fully discharged, or discharged to the point where operation is no longer practical, the word "BATTERY" flashes, and the camcorder shuts off.

The *tape/counter memory indication* is not recorded on tape, but shows the amount of tape available. The letter "M" indicates that the memory

Sec. 2-4   Typical Operating Procedures   45

function is operating (the camcorder stops when a tape count of 0000 is reached).

The *date indication* is recorded on tape, along with whatever scene is being recorded by the camera, but can be set only during the RECORD/PAUSE mode.

The *recording indication* (a green light at the bottom of the EVF screen) turns on when a recording is being made.

The *in/outdoor indication* (an orange light at the top of the EVF screen) turns on when the outdoor setting is selected.

When the *low-light indication* (the word "LIGHT" in the center of the EVF screen) appears, the light level is inadequate for good recording. Note that not all camcorders have a low-light indicator.

Typically, the low-light indicator circuit involves monitoring the video signal (or possibly the iris control signal) for some minimum value. If the light and video signal are below the desired limits, the system control produces a signal that places a warning on the EVF screen (through the same character generator that produces the other warning signals). In some camcorders, a low-light condition turns on a lamp next to the EVF screen.

When the *tape warning indication* (the word "TAPE" in the center of the EVF screen) flashes, the *cassette tab* is missing (the cassette is not to be re-recorded), and recording operation does not start. This is discussed further in Sec. 2-4.2.

The *dew indication* (the word "DEW" in the center of the EVF screen) appears if excessive moisture condenses in the camcorder. As in the case of any VCR, damage can result if the tape is driven under conditions of excessive moisture. When the dew indication is present, the camcorder is made inoperative (the dew indicator flashes for about 10 sec, and the camcorder is shut off). In this case, wait until the dew indicator no longer flashes when the camcorder is turned on again.

Note that when more than one warning is required to be on display in the EVF, the words "BATTERY" and "TAPE" flash alternately.

2-4.2 Information on Cassettes

The cassettes used in VHS camcorders are identical with the cassettes used in VHS VCRs and are completely interchangeable. The following notes are included on the offchance that you do not know all about VHS cassettes. Figure 2-12 summarizes the most important points.

Use only VHS cassettes (not Beta or 8 mm) in a VHS camcorder. As in the case of a VCR operated in the SP mode, the record/playback times for standard VHS cassettes used in a camcorder are 2 hr and 40 min for T-160, 2 hr for T-120, and 1 hr for T-60.

Videocassettes and their tapes are made with high precision, so be sure to observe the following:

46 User Controls, Operation Procedures, Accessories, and Interconnections   Chap. 2

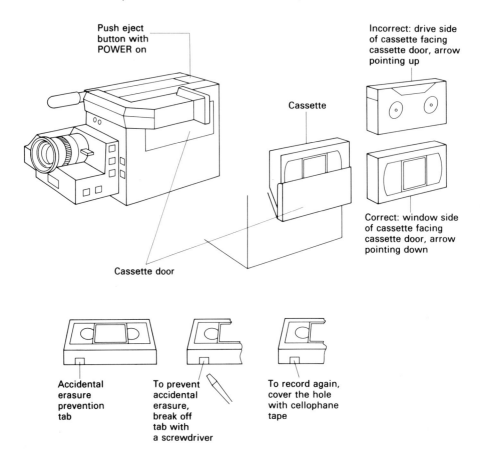

FIGURE 2-12   Information on cassettes

Do not subject the tape to unnecessary movement.

Do not insert and eject the cassette over and over again. This may result in tape looseness or scratches on the tape.

Never try to disassemble the cassette or splice the tape.

Do not open the cassette, touch the tape, or put anything inside the cassette.

Do not place cassettes in direct sunlight, and keep cassettes away from any source of heat.

Do not drop the cassette. Do not subject cassettes to intense shaking or vibration.

Store cassettes vertically, and avoid storage where humidity is high, where there is dust, and where there are magnetic fields (near magnets, motors, transformers, loudspeakers, etc.).

Accidental erasure of recorded material can be prevented by breaking

Sec. 2-4   Typical Operating Procedures   47

off the tab on the rear of the cassette with a screwdriver or similar tool, as shown in Fig. 2-12. This causes the erasure prevention mechanism to function, as is the case with a VCR. Typically, with the tab removed, a switch is actuated (or prevented from being actuated). This signals the system control to prevent record operation under any conditions.

To record on a cassette that has been protected by removal of the tab, cover the hole (where the tab was) with cellophane tape.

Figure 2-12 shows the steps to insert a cassette in the camcorder shown in Figs. 2-3 through 2-5. The procedure is essentially the same (but not necessarily identical) for most VHS camcorders.

First, press the *power button*, and then press the *eject button* to open the cassette compartment door.

Insert the cassette and close the cassette door *gently*.

There is probably something wrong if you must force the door closed. For example, there may be something in the cassette compartment. More likely, the cassette has been inserted upside down, or backward, or both. As shown in Fig. 2-12, the cassette should be inserted with the window side of the cassette facing the door, and the arrow pointing down. Do not attempt to insert a cassette that is upside down or backward (especially if the customer is watching!).

2-4.3 The Basic Record/Playback Sequence

Before starting any record/playback operation, make certain that the camcorder is properly connected to a power source or battery and that a known-good TV is available for playback. Then proceed as follows:

1. Push the power switch on. The power indicator next to the power button should light.
2. Adjust the EVF to your individual vision, and check the EVF for any warning signals (BATTERY, LIGHT, TAPE, DEW). Also check that there are four dashes between the E and F displays (Fig. 2-11) to indicate that the battery is fully charged and that the counter shows 0000. If necessary, reset the counter with the reset button, as discussed in Sec. 2-4.2.

    If the BATTERY signal is flashing, check the battery charge. If the TAPE signal is flashing, this is normal if there is no cassette installed or if the cassette has the erasure tab removed. If the DEW signal flashes for about 10 secs, and the camcorder remains off (no operating mode can be selected), wait until the moisture problem is cleared. If the LIGHT signal is on, check that the *lens cap is removed* (this often works wonders!). If necessary, provide more light.
3. Push the eject button and insert a cassette (with the erase tab, or cellophane tape, in place) into the cassette compartment.

4. The camcorder should go into the record/pause mode automatically when power is first applied (and the cassette is installed). The record/play and pause/still indicators on the corresponding buttons should go on.
5. Set the in/outdoor switch to the appropriate position. Note that not all camcorders have an in/outdoor select function. (You are supposed to get proper white balance when the white balance is set to automatic.)
6. Set both the white-balance and focus switches to automatic. Note that on some camcorders, you must also set an iris switch to automatic.
7. Press record/pause to start recording. The record indicator in the EVF should turn on. To pause when recording, press record/pause again. Do not confuse the record/pause button (Fig. 2-3) with the record/play select or pause/still buttons (Fig. 2-4). Also, do not confuse the record/play select indicator (Fig. 2-4) with the record indicator in the EVF (Fig. 2-11).
8. If you plan on a long pause, push standby, and check that the standby indicator turns on. This conserves energy, but causes the picture to removed from the EVF. Always go to STANDBY from the PAUSE mode, not from RECORD.
9. Connect an earphone to the earphone jack if you want to monitor the sound track during record.
10. When recording is complete and/or you are ready for playback, push stop. This releases the PAUSE mode. Note that if the STOP mode continues for more than 5 min (when battery power is used), the camcorder goes to standby.
11. Press rewind and wait until the tape rewinds completely. The EVF tape count should show 0000. If necessary, reset the tape counter. Refer to Sec. 2-4.4.
12. Press play. The play indicator should turn on, and the recorded picture should appear on the EVF after a few seconds. If desired, you can also play back the recording on a TV as described in Sec. 2-4.14.

Keep the following points in mind for the normal record/play sequence.

1. If necessary, adjust the tracking control to remove noise bars in the picture during playback. Normally, the tracking control should be left in the fixed or detent position when playing back tapes recorded on the *same camcorder.*
2. When both the record/play select and pause/still indicators are on steady (not flashing), the camcorder is in the RECORD/PAUSE mode. You can then toggle between RECORD and PAUSE with the record/pause button.

Sec. 2-4   Typical Operating Procedures   49

3. If the record/play select indicator is flashing, this usually indicates that you have installed a cassette with the erase tab removed (or there is no cassette installed). Under these conditions, the camcorder should be in the STOP mode.
4. If the cassette tab is in place, the camcorder should go directly into RECORD/PAUSE mode when power is first applied.
5. The camcorder goes into the PLAY mode from either of two conditions. First, if the record/play select indicator is off, the camcorder goes to PLAY mode when stop is pressed. If the record/play select indicator is on, push record/play select, and then push stop. Either way, playback then starts when play is pushed.
6. To return to the RECORD mode from PLAY, push record/play select again, and check that the record/play select indicator turns on.
7. After either record or playback is complete, make a habit of fully rewinding the tape by pushing rewind before pushing eject.
8. When the tape reaches the end during playback, or fast forward, the tape rewinds automatically.
9. If you play tapes recorded in LP or SLP (on a VCR), the display (on both the EVF and TV) will probably be snowy.

2-4.4 Using the Tape Counter

The tape counter keeps accurate track of where the tape is at all times while in play, record, special-effects playback, pause, fast forward, or rewind. When the memory function is selected (with the counter at 0000), the camcorder stops during rewind or fast forward whenever the tape counter reaches 0000. The normal tape counter operating sequence is as follows.

1. Push reset so that the tape counter in the EVF is set to 0000. Note that the tape counter is not always in the EVF on all camcorders. For example, see the camcorder of Fig. 2-2. However, the tape counters of virtually all camcorders have a reset and memory function.
2. To use the memory feature, push memory/display. The letter M should appear on the EVF screen to produce a display of M0000.
3. After recording or playback is complete, push stop. The tape count should stop simultaneously.
4. Push rewind or fast forward. The tape should stop whenever M0000 is reached.
5. On camcorders where the tape count and other displays (such as battery charge) are in the EVF, it is possible to clear all such displays from the EVF to get an unobstructed view of the scene by pushing the memory/display button repeatedly. The sequence is as follows.

If the tape-count and battery-charge displays are present, and the tape count does not show an M, push memory/display once to produce the M, and then push memory/display again to remove both the tape-count and battery-charge displays.

If the scene is present, but there are no displays, push memory/display once to restore the tape-count and battery-charge displays. Push memory/display again, to produce the M in front of the tape-count display (M0000).

If the battery-charge display is present, and the tape count shows an M, push memory/display once to remove both displays.

2-4.5 Using the Record/Review Feature

The record/review function allows you to monitor the last few seconds of recorded material on the EVF screen (or TV) by pressing record/review when the camcorder is in the RECORD/PAUSE mode (record/play select and pause/still indicators on steady). When this brief playback is complete, the camcorder returns to the RECORD/PAUSE mode automatically, and the next segment is recorded in synchronization with the last segment recorded.

2-4.6 Using Fine Edit for Proper Continuity

Here is a tip you can pass on to your customers who want proper continuity when taping from the STOP mode or after changing the battery. If you simply go back to the normal record/play sequence after a full stop (not a pause), you may produce gaps in the tape.

1. Rewind the tape for a short distance (about five digits on the tape counter) by pushing rewind.
2. Make certain that the record/play select indicator is off (push record/play select if necessary) and that the pause/still indicator is on (push press/still if necessary).
3. Push play and view the scene on the EVF.
4. At the location where you wish to continue recording, push pause.
5. Push record/play select, and check that the record/play select indicator turns on (indicating that you are in the record/pause mode).
6. Push record/pause to continue recording.

2-4.7 Using the Focus System

Virtually all camcorders can be focused both manually and automatically. The camcorder automatically focuses on the subject placed in the center of the EVF when the focus switch is set to auto or automatic. For manual focus, set the focus to manual, and adjust the lens focus ring for proper focus.

Pressing push-to-focus (while in the manual position) turns on the autofocus system, until push-to-focus is released. Should the subject change, refocusing may be required. In any event, the focus setting will not automatically change until push-to-focus is pushed again.

Keep in mind that the autofocus system will not work if the distance detection window (or autofocus window) is covered, even partially. Also, the autofocus system may not focus on an object over about 40 feet from the camcorder. Likewise, if the object being shot is small, autofocus operation may be erratic. The autofocus system needs a reflected area for proper operation. As a guideline, when camcorder-to-object distance is 10 ft, an object must have a reflected area greater than 4 in. in diameter. Keep all these points in mind when troubleshooting a "the autofocus system doesn't work" symptom!

Generally, manual focus is recommended primarily where the scene has both far and near objects (such as in a scene where there is considerable perspective), in scenes with fine horizontal patterns, and where the object does not reflect light (such as a nonreflective black curtain). Keep in mind that camcorder autofocus systems operate on reflected infrared (IR) light, not on sonic waves as do some still cameras.

Here are some tips you can pass on to your customers who want to use manual focus effectively.

1. Aim the camcorder at the scene to be recorded.
2. Press and hold the T-side of the power zoom switch until the lens zooms to maximum close-up setting. Refer to Sec. 2-4.9.
3. Rotate the lens focus ring for the sharpest picture at this setting. (It may be necessary to back away from the subject to focus the image properly.)
4. Press the W-side of the power zoom switch. The picture should stay in focus over the entire zoom range. However, refocusing may be required when the camcorder is aimed at a new scene.

2-4.8 Using the White-Balance System

Virtually all camcorders have an automatic white-balance system to maintain the correct color balance under a wide range of lighting conditions. Likewise, camcorders have some form of manual control for special lighting conditions.

On camcorders such as shown in Fig. 2-2, set white balance to AUTO for full automatic operation. For special lighting, hold white balance in the AUTO SET position until the record/battery indicator (in the EVF) stops flashing. This indicates that white balance is properly set for the special lighting condition and will remain until reset.

On camcorders such as those shown in Figs. 2-3 through 2-5, first set the in/outdoor switch as required, and then set white balance to AUTO for full automatic operation. Make sure that the color-control dial is set at the fixed or detent position. For special lighting conditions, set white balance to FIXED, and then use the color-control dial to adjust for special lighting conditions (counterclockwise for blue, clockwise for red).

No matter what controls are involved, keep the following points in mind when using the white-balance system.

Normally, white balance should be set to automatic for best results. Manual white balance is recommended for scenes of a single predominant color. For example, use manual when an object is a primary color such as green, red, or blue; when shooting at sunset or sunrise; or when one color (any color) dominates the scene.

One common operating technique for adjusting white balance is to monitor the scene on a TV, and then adjust the manual white-balance controls for best results. If you use this technique, make certain that the color controls on the TV are properly set (or play back all recordings only with that TV!).

2-4.9 Using the Zoom Lens

Virtually all camcorders have both manual and power (motor-driven) zoom lenses. Figure 2-13 shows the controls involved.

You can manually zoom at any time. Typically, the zoom ring has a lever for steady turning, as shown. On most camcorders, the power zoom switch is conveniently located at the fingertips on the hand grip. This makes it possible to control the power zoom with one hand, while focusing the lens with the other (using the lens focus ring).

Power zoom is selected by lightly pushing the T (telephoto) or W (wide-angle) sides of the power zoom switch. As shown in Fig. 2-13, the lens moves in or out, and the zoom ring lever rotates accordingly.

2-4.10 Taking Close-up Pictures

The zoom feature makes it possible to take close-up pictures (typically as close as 1 in. from the lens). The following procedure applies to most camcorders.

1. Rotate the zoom ring lever to the extreme left position (wide angle).
2. Push the macro button on the zoom lever (Fig. 2-13) while moving the lever further to the left.
3. Set the focus switch to manual (Sec. 2-4.7).
4. Now use the zoom ring lever for macro focus adjustment. Note that

Sec. 2-4   Typical Operating Procedures

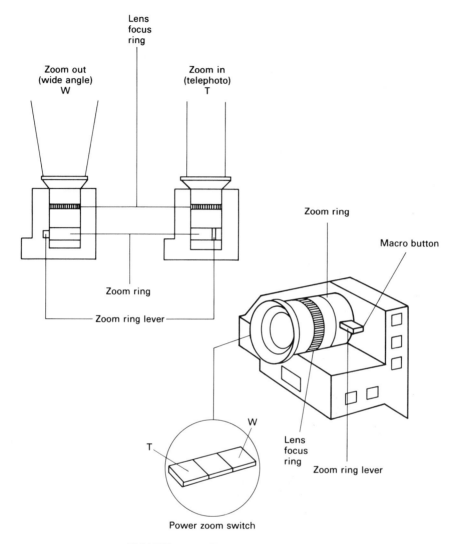

FIGURE 2-13 Typical zoom controls

this macro range is to be used for close-up pictures only, and not as an extreme wide-angle range. On most camcorders, the macro range produces blurred images for subjects not close up.

2-4.11 Using the Backlight

Camcorders such as those shown in Figs. 2-3 through 2-5 have a backlight adjustment feature (not found on all camcorders). The backlight feature is used to adjust the light level on a subject that is darker than the surrounding scene. The backlight adjustment is useful for recording a subject in a shadowed area or in front of a strong backlight. Without a backlight adjustment, subjects recorded in front of a strong backlight often appear as silhouettes.

Simply push the backlight button while recording high-contrast or strong backlight scenes. Release the button when the lighting conditions return to normal.

On camcorders such as those shown in Fig. 2-2 (that do not have a backlight control), it is sometimes possible to get the same effect with the manual iris control. However, as discussed in Sec. 2-5, camcorders should *never be pointed at a bright light source*, particularly with the iris open. The strong light can damage the pickup.

2-4.12 Recording the Date

On most camcorders, the date display (month, day, year) can be set in the RECORD/PAUSE or STOP modes, and then recorded as an overlay in recording (when the display is on). The following procedure applies specifically to camcorders such as shown in Figs. 2-3 through 2-5, but are typical for many other camcorders.

1. Push the date button so that 00.00.00 appears on the EVF screen. The date display disappears when the date button is pressed again.
2. Push the set button to select the date. Holding the set button causes the numbers to cycle. That is, the digits cycle from 0 to 1, 1 to 2, and so on as the set button is held.
3. Push the select button to select which digits are to be cycled (month, day, year) by means of a cursor bar (that does not appear in the recorded overlay).
4. If you do not want the selected date to be recorded at some point on the tape, push the date button so that the date display disappears. Push the date button again to record the date on other portions of the tape.

Note that if power is turned off for more than 30 min, the date display returns to 00.00.00 (on most camcorders).

2-4.13 Using the Remote Control

The remote control feature found on many camcorders makes it possible to record from a remote location without disturbing the camcorder or the subject. Refer to Sec. 2-3.4. Simply connect the remote control cable to the corresponding remote jack on the camcorder. Then set the camcorder in the REMOTE/PAUSE mode. Under these conditions, pushing the remote record button starts a recording. Pushing the remote record button again

Sec. 2-4   Typical Operating Procedures   55

resets the camcorder to pause. On most camcorder remote controls, a light turns on when the camcorder is in record and turns off when PAUSE is selected.

Keep in mind that all other controls on the camcorder must be preset for proper light, focus, and so on. The remote unit controls only the record/pause selection function.

2-4.14 Playback with a TV

Camcorders can be used to play back recordings on a TV, thus eliminating the need for a VCR. The following procedure applies specifically to camcorders such as shown in Figs. 2-3 through 2-5, but are typical for many other camcorders.

1. Connect the camcorder to the TV as described in Sec. 2-3.2.
2. Push power. Then push eject and install a cassette.
3. Push stop to release the PAUSE mode. Then push record/play select, and check that the corresponding indicator turns off to indicate that the camcorder is in play.
4. Push rewind as necessary to rewind the tape completely. Reset the tape counter if necessary.
5. Push play. The recorded picture should appear on the TV screen after a few seconds. Note that if the tape was originally recorded in LP or SLP, the playback may be snowy. Also, it may be necessary to use the tracking control to eliminate noise bars in the TV picture, if the tape was recorded on another camcorder or a VCR.
6. If you want to return to the record mode, push record/play select, and check that the corresponding indicator turns on to indicate that the camcorder is in record. Keep in mind that if you record on a tape with the erase tab in place, you will erase the original recording!

2-4.15 Using Special Effects during Playback

Camcorders such as those shown in Figs. 2-3 through 2-5 have two special effects (that can be used during playback only): forward and reverse *search* to locate a particular segment on the tape rapidly and *still* playback to view a still picture.

To use search, push remote/play select as necessary to place the camcorder in the PLAY mode (remote/play select indicator off). Then push play and fast forward/search or rewind/search as required. Search speed is about three times normal speed. During search, horizontal noise bars will appear on the TV screen (and EVF) and there will be no audio.

To view a still picture during playback, press pause/still as necessary until a still picture appears (pause/still indicator on). Press pause/still again to release the PAUSE/STILL mode.

2-5 PRECAUTIONS

Although designed for portable, outdoor use, a camcorder is still a delicate electronic instrument and should be so treated. Here are some precautions to be considered for camcorders, in addition to the usual precautions for any portable electronic device or camera.

Always transport the camcorder in an accessory carrying case, if possible. Such cases are usually padded to minimize shock (as when tossed into the trunk compartment of an auto).

A camcorder has been designed for outdoor use in all kinds of weather. However, camcorders will probably not survive any type of direct exposure to water, rain, sleet, snow, or dark of night, or a direct splashing from a pool, or even a cup of coffee.

To avoid shock hazard when operated from line power, do not operate the camcorder if exposed to rain or moisture. Never operate the camcorder from line power if the camcorder and/or adapter are wet (from any cause).

Unplug the camcorder from the line power when not in use, particularly when cleaning the camcorder.

Be sure to observe the following:

1. *Do not aim the camcorder at the sun or other bright objects.* This can permanently damage the pickup (either tube or solid state), even if the camcorder is turned off and disconnected from any power.
2. *Do not leave the camcorder in direct sunlight.* When the EVF eyepiece is exposed to direct sunlight, the eyepiece lens acts as a magnifying glass. The concentrated sunlight can cause damage to internal parts of the camcorder.
3. *Avoid sudden changes in temperature.* If the camcorder is suddenly moved from a cold place to a warm place, moisture may form on the tape and inside the camcorder. In this case, the dew indicator flashes, and power is shut off automatically (on most camcorders).
4. *Do not leave the camcorder power turned on*, even when you intend using the camcorder in a short time. Typically, a long warm-up time is not required for normal operation (although you may want a longer warm-up period when making adjustments). Most camcorders are ready for normal operation in seconds.

If you must keep the power on so that the camcorder is available for immediate use, put the camcorder in the standby mode, after pause is selected. In that way, you can go to record or play instantly. On many camcorders,

Sec. 2-5 Precautions

the circuits go into a standby condition (or possibly a full stop) after a short time (typically 5 min) after pause is selected.

Store and handle the camcorder for a minimum of unnecessary movement (avoid striking or shaking the camcorder, obviously). The pickup tube is generally the most sensitive part of the camcorder, with the EVF tube running a close second.

Do not use strong or abrasive detergents when cleaning the camcorder body and outer surfaces. Refer to Sec. 3-5 for cleaning information.

To protect the lens, always replace the lens cap when the camcorder is not in use. Do not touch the lens surface. Use a commercial lens solution and/or lens paper when cleaning the lens. Improper cleaning can scratch the lens coating.

Be careful when connecting the RF adapter to a TV set. Make the connections only as shown in Figs. 2-7 and 2-8 or as specified in the operating instructions for the particular camcorder. Failure to do so may result in operation that violates Federal Communication Commission (FCC) regulations regarding the use and operation of RF devices. (You may broadcast camcorder home movies to the entire neighborhood! While this may be amusing, it is also illegal.) *Never connect* the output of the RF adapter to an antenna or make simultaneous (parallel) antenna and adapter connections at the TV antenna terminals.

If the RF adapter and/or camcorder do cause interference to radio or TV reception, with proper connections, here are some tips to minimize the problem.

First, confirm that the interference is being caused by the camcorder in question. (Turn the camcorder on and off.) Then, try reorienting the antennas and/or moving the camcorder in relation to the TV. The lead-in may be radiating the undesired signal. Finally, try plugging the camcorder and/or TV into a different power outlet, or operate the camcorder with battery power. Try a substitute RF adapter, if practical.

If none of these suggestions works, you may find the following booklet prepared by the FCC helpful: *How to Identify and Resolve Radio-TV Interference Problems*, available from the U.S. Government Printing Office, Washington, D.C. 20402, Stock No. 004-000-00345-4.

3

TEST EQUIPMENT, TOOLS, and ROUTINE MAINTENANCE

This chapter describes the test equipment and tools you will need for camcorder service. We also discuss routine maintenance for camcorders. Keep in mind that the information in this chapter is general. If you are going to service a particular camcorder, get all the service information you can on that camcorder. Likewise, if you plan to go into camcorder service on a large scale, study all the applicable service literature you can find (then, when all else fails, you can follow instructions). We discuss adjustments (both mechanical and electrical) using the tools and test equipment in Chapter 8. We also discuss use of the tools and test equipment for troubleshooting in Chapter 9.

3-1 SAFETY PRECAUTIONS IN CAMCORDER SERVICE

In addition to a routine operating procedure (for both test equipment and the camcorder), certain precautions must be observed during operation of any electronic test equipment. It is assumed that you are already familiar with TV/VCR service precautions and procedures (both color and black and white), including the use of test equipment, safety precautions (leakage current checks, handling electrostatically sensitive or ES devices, leadless component removal, high-voltage checks at picture tubes, etc.), installation, routine maintenance, isolation transformers, operating procedures, and basic troubleshooting (including solid-state troubleshooting).

If you are not familiar with any of these basic precautions and procedures, and you plan to service camcorders, you are in terrible trouble. You had better read the author's best selling *Handbook of Simplified Television Service* (1977),[1] *Handbook of Advanced Troubleshooting* (1983), *Complete Guide to Videocassette Recorder Operation and Service* (1983), and *Complete Guide to Modern VCR Troubleshooting and Repair* (1985).

3-1.1 Copyright Problems

Many of the programs broadcast by television stations and cable networks are protected by copyright, and federal law imposes strict penalties for copyright infringement. Some motion picture companies have taken the position that home recording for noncommercial purposes is an infringement of their copyrights. The U.S. Supreme Court has ruled otherwise. However, there are laws before the U.S. Congress to protect unauthorized recording of TV broadcast material. Likewise, the recording—or even the viewing—of cable broadcasts without authorization (without paying for the cable service) is against the law. So until all these laws have been sorted out, a camcorder used to record copyrighted material (on tape or from a TV screen) should be operated at the user's own risk.

3-2 TEST EQUIPMENT FOR CAMCORDER SERVICE

Much of the test equipment used in camcorder service is basically the same as that used in TV/VCR service. That is, most service procedures are performed using meters, signal generators, NTSC color generators, oscilloscopes, frequency counters, waveform monitors, vectorscopes, power supplies, assorted patch cords, and so on.

As a practical matter, the service procedures for the VCR section of a camcorder can be performed using conventional test equipment (found in a typical TV/VCR service shop), provided that the oscilloscopes have the necessary gain and bandpass characteristics, that the signal generators cover the appropriate frequencies, and so on. For these reasons, we do not go into full details on basic test equipment, except to say that the *oscilloscope should have a delayed sweep*, and *25-MHz bandwidth*, with *channel-invert capability*. If you have a good set of test equipment suitable for conventional TV/VCR work, you can probably service the VCR portion of any camcorder.

On the other hand, the camera section of a camcorder requires specialized test equipment, particularly for adjustment. Figure 3-1 illustrates a typical setup for camcorder camera adjustment. The following paragraphs discuss the features for such equipment.

---

[1]All titles have been published by Prentice-Hall, Inc., Englewood Cliffs, N.J.

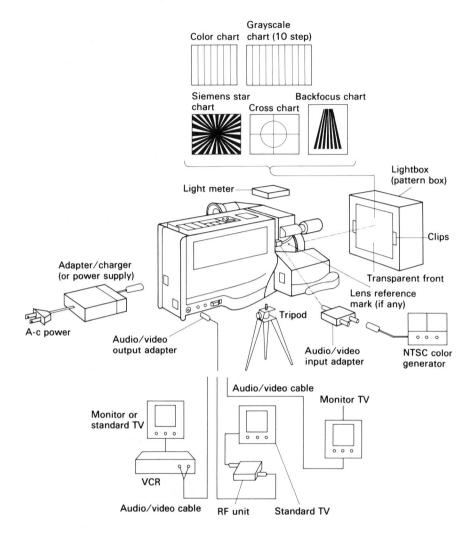

FIGURE 3-1 Typical setup for camcorder camera adjustments

3-2.1 Tripod

Obviously, the camcorder must be held steady during camera adjustment. A tripod is your best bet, although some technicians prefer to mount the camcorder on a bench. Either way, make certain that the tripod or mount matches the mounting hole on the camcorder bottom (Fig. 2-3).

Also check if the camcorder has a *lens reference mark*. For many adjustments, the camera must be mounted a precise distance from the object being photographed. The lens reference mark (if any) is usually located on top of the lens and is used as a reference point for measurement to the object. In most cases, the service literature simply specifies a distance from the front end of the lens.

3-2.2 Light Source and Light Meter

Many adjustments required that the object be photographed using a particular light source. Typically, the light source must have a color temperature of 3200°K. (A 3200°K quartz lamp is often recommended.) No matter what the source, use the *correct color temperature* when making any camera adjustments. If not, the colors will never be quite right!

Any light meter can be used. Preferably, the *meter should read in lux*, rather than footcandles, since most service literature specifies lux. For example, a typical camera adjustment procedure specifies 100 lux reflected from an object (a grayscale chart), 6 ft away from the lens reference mark.

If your light meter reads only in footcandles, remember that 10 lux is *approximately* equal to 1 footcandle.

3-2.3 Lightbox or Pattern Box

A lightbox (also known as a pattern box) is essentially a light source (typically a 3200°K quartz lamp producing about 100 lux) within a box. The front of the box is transparent (Fig. 3-1) and has clips or slots to hold various adjustment charts (Sec. 3-2.4).

An alternate for the lightbox is to mount the charts on a flat surface (at the specified distance from the camera lens) and then reflect the correct light from the charts onto the lens.

3-2.4 Adjustment Charts

Although there is no standardization, a typical set of adjustment charts includes a *grayscale chart*, an NTSC *color chart*, an *autofocus chart*, and a *backfocus chart*.

The grayscale chart provides for black and white adjustment, while the NTSC chart provides for color adjustments. The various other charts shown in Fig. 3-1 (*Siemens star*, *backfocus*, *cross*, etc.) provide a fixed reference for both check and adjustment of the camera.

3-2.5 Monitor TV

The monitor-type TV sets designed specifically for VCRs, videodisc players, and video games provide the most practical means of monitoring a camcorder, both during and after adjustment/troubleshooting. However, if you are planning to go into camcorder service on a grand scale, you may want to consider a *receiver/monitor* such as used in *studio* or *industrial video* work. These receiver/monitors are essentially TV receivers, but with video and audio *inputs and outputs* brought out to some accessible point (typically the front panel).

The output connections (not found on a monitor TV) make it possible to monitor broadcast video and audio signals as they appear at the output of a TV IF section (the so-called *baseband signals*, generally in the range of 0 to 4.5 MHz, at 1 V peak to peak for video and 0 dB, or 0.775 V, for audio). The output signals from the receiver/monitor can be injected into the camcorder.

The input connections on either a receiver/monitor or monitor-type TV make it possible to inject video and audio signals from the camcorder (without using an RF unit) and monitor the display. Thus, the baseband output from the camcorder can be checked independently from the RF unit. When an RF unit is used, it is often difficult to tell if faults are present in the camcorder audio/video or in the RF unit.

If you use a TV set as a monitor, some technicians recommend that the vertical height control be adjusted to underscan the picture. This makes it easier to see the video switching point in relation to the start of vertical blanking.

No matter what is used as the monitor, make certain that the *colors are properly adjusted*! If you adjust a camcorder to produce good colors on an improperly adjusted monitor, you are in trouble!

3-2.6 Digital Test Equipment

Camcorders use some form of microprocessor in the system-control circuits. Often, more than one microprocessor is used throughout the camcorder circuits. While it is possible to monitor most microprocessor signals with an oscilloscope, there are special test instruments for digital troubleshooting.

The most useful tools for monitoring the digital signals associated with microprocessors in camcorders include the *logic probe* and, possibly, the *logic pulser* and *current tracer*. *Logic comparators, logic analyzers*, and *signature analyzers* are of little value in camcorder service.

We do not go into the use of digital test equipment in this book. To do so requires too much space. If you are not familiar with digital troubleshooting techniques, your attention is directed to the author's best-selling *Handbook of Advanced Troubleshooting*.[2]

[2]Englewood Cliffs, N. J.; Prentice-Hall, Inc., 1983.

3-3 TOOLS AND FIXTURES FOR CAMCORDER SERVICE

Figure 3-2 shows some typical tools and fixtures for field service of camcorders. These tools are available from the camcorder manufacturer. In some cases, complete tool kits are made available. Those readers already familiar with VCR service will recognize most of these tools, or their equivalents. (If you are not familiar with VCR service, you had better learn before you attack a camcorder!) So we do not dwell on tools and fixtures in this book. However, here are some thoughts to consider.

There are factory tools and fixtures used by the manufacturer for both assembly and service of camcorders. These factory tools are not available for field service (not even to factory service centers, in some cases). This is the manufacturer's subtle way of telling service technicians that they should not attempt any adjustments, electrical or mechanical, not recommended in the service literature. Take this subtle hint!

The author has heard many horror stories from factory service people concerning "all hope gone" camcorders brought in from the field. Most of the problems are the result of tinkering with mechanical adjustment (although there are some technicians who can kill a camcorder with a simple electrical adjustment). To avoid this problem, use only recommended tools and perform only recommended adjustment procedures.

Most camcorder manufacturers provide an *alignment tape* (or set of tapes) as part of their recommended tools. As shown in Fig. 3-2, an alignment tape is housed within a standard cassette and has several very useful signals recorded at the factory using very precise test equipment and signal sources.

Although there is no standardization, a typical set of alignment tapes might include a tape with a black and white pattern and a 7-kHz mono-audio signal, a tape with a color pattern (bars) and a 1-kHz mono-audio signal, and a tape with a multiburst pattern and a 3-kHz stereo-audio signal. If you intend to service one type of camcorder extensively, you will do well to invest in the recommended tapes.

A typical use for the signals recorded on an alignment tape is to check overall operation of servo speed-control and phase-control systems. For example, if the frequency of an audio playback is exactly the same as that recorded (or within a given tolerance), and remains so for the entire audio portion of the tape, as checked on a frequency counter, the servo-control systems, both speed and phase, must be functioning normally. If there are any mechanical variations or variations in servo control that produce wow, flutter, jitter, and so on, the audio playback varies from the recorded frequency.

If you do not want to invest in factory alignment tapes, or if you do not want to wear out an expensive factory tape for routine adjustments (alignment

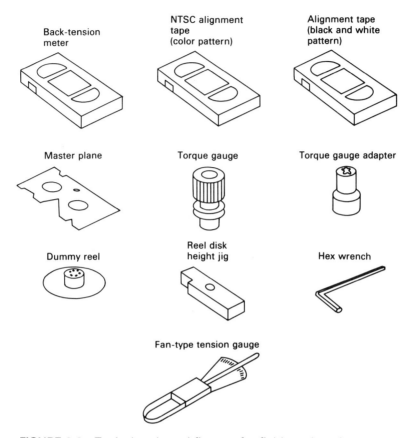

FIGURE 3-2 Typical tools and fixtures for field service of camcorders

tape deteriorates with continued use), you can make up your own alignment tapes or "work tapes" using blank cassettes. The TV and cable stations in most areas broadcast color bars before or after regular programming. These colorbars can be recorded using a camcorder or VCR known to be in good operating condition. A *stationary color pattern with vertical lines* is especially useful. If you have access to a factory tape, you can duplicate the tape on your own work tape (making certain to use a known-good camcorder/VCR when making the duplication).

In addition to the special tools described thus far, the mechanical sections of most camcorders can be disassembled, adjusted, and reassembled with common handtools such as wrenches and screwdrivers. Keep in mind that most camcorders are manufactured to Japanese *metric standards* and your tools must match. For example, you will need metric-sized Allen wrenches and Phillips screwdrivers with the Japanese metric points.

Since camcorders require periodic cleaning and lubrication, you will also

Sec. 3-4   Routine Maintenance for Camcorders   65

need tools and applicators to apply the solvents and lubricants (cleaner sticks for the heads, etc.). Always use the recommended cleaners, lubrications, and applicators, as discussed in Sec. 3-4.

3-4 ROUTINE MAINTENANCE FOR CAMCORDERS

There is considerable disagreement among camcorder manufacturers concerning the need for routine checks or periodic maintenance. At one extreme, a certain manufacturer recommends replacement of a few parts after a given number of playing hours or playing times. They recommend that all motors be replaced after 10,000 hours (who counts?). At the other extreme, another manufacturer recommends no routine maintenance for the first year of camcorder use, and then some cleaning/lubrication when the camcorder is in for service. "Fix it if it breaks down" is the rule. (It is fair to say that this rule will be observed religiously.)

Somewhere between these two extremes, other manufacturers recommend adjustment (electrical and/or mechanical) only as needed to put the camcorder back in service or when certain parts or assemblies have been replaced. However, most camcorder manufacturers recommend a complete checkout, using a known-good monitor, after any service.

The author has no recommendations in the area of routine maintenance, except that you follow the manufacturer's recommendations. The author also realizes that the general public regards a camcorder in the same way it does a TV or VCR (that is, "Bring it in for service when it breaks down"). However, here are some procedures that can be applied when a camcorder is brought to you for service.

3-4.1 Cleaning

To clean the outside of a camcorder, use a soft, clean cloth to wipe off dust and accumulated dirt. If absolutely necessary, moisten a soft cloth with *diluted* neutral detergent to remove heavy dirt. *Never use* paint thinner, benzene, or other solvents, since any of these react with the surface and cause color changes and possible melting.

The lens requires particular care in cleaning. The lens surface must be clean (and free of moisture) for proper operation of a camcorder. Try not to touch the lens surface, even if your fingers are clean. Body oils can leave smudges on the lens. Keep the lens cap on (except when the cap must be off for service and/or adjustment). Dust can be removed from the lens with an air blower (designed for use on cameras). Dirt can be removed with camera lens cleaners (cleaning papers).

There are solvents and cleaning pads (such as Kimwipes) to clean the mechanical sections of the camcorder (the VCR portion). Likewise, there are head-cleaning kits and spray cans of head cleaners for the video heads, full-erase head, and audio-control heads. The kits usually contain cleaning sticks or wands.

Instead of solvents or cleaners, most manufacturers will go along with the use of alcohol for all cleaning. Methyl alcohol does the best cleaning job, but it can be a health hazard (especially if you drink it!). Isopropyl alcohol is usually satisfactory for most cleaning (do not drink it either).

We do not go into cleaning in any detail here. This is one area where the instructions found in camcorder service manuals are usually good. Follow the instructions! In the absence of any instructions, and as a general procedure for all camcorders brought in for service, use the following to clean the heads and tape path.

1. Be sure to turn off the power and/or remove the battery before you start cleaning the VCR section mechanism.
2. Rotate the video head cylinder (or drum, or scanner, whichever term you prefer) to a position convenient for cleaning the video head.
3. Moisten a cleaner stick or wand with alcohol (or a cleaner approved by the manufacturer). Lightly press the buckskin portion of the stick against the video head drum, and move the head by turning the drum back and forth. Clean all four video heads in the same way.
4. *Always pass the stick across the head in the same direction as the tape path.* For example, if the tape moves horizontally across the heads, *never* move the cleaning stick vertically across the head. Cleaning across the tape path can damage the heads.
5. To clean the full-erase and audio-control heads, moisten the cleaner stick with alcohol or cleaner, press the stick against each head surface, and clean the heads by moving the stick in the same direction as the tape path.
6. Clean the drum surface, and each surface where the tape passes (tape guides, etc.) with a soft cloth moistened with alcohol or cleaner. When cleaning the drum surface, be careful not to touch the video heads with the cleaning cloth. If necessary, rotate the drum to move the head away from the spot to be cleaned.

Finally, no discussion of VCR/camcorder cleaning is complete without some mention of *cleaning cassettes*, also known as *lapping cassettes*. Such cassettes contain a nonmagnetic tape coated with an abrasive. The idea is to load the lapping cassette and run the abrasive tape through the normal tape path (across the video heads, around tape guides, etc.) *for a few seconds*. This cleans the entire tape path (especially the video heads) quite thoroughly.

Prolonged use of a lapping tape can result in permanent damage (especially to the video heads). The author has no recommendation regarding cleaning tape. If you decide on a lapping tape, always follow the manufacturer's recommendations, and *never use any cleaning tape for more than a few seconds.*

3-4.2 Lubrication

Never lubricate or clean any part not recommended by the manufacturer. Most camcorders use sealed bearings that do not require lubrication. A drop or two of oil in the wrong places can cause damage. When relubricating, remove the old lubricant first; then sparingly apply new lubricant. Clean off any excess or spilled oil. In the absence of any recommendations, use a light machine oil such as sewing machine oil. Fortunately, as in the case of cleaning instructions, the lubrication sections of most camcorder service literature are well written.

4

CAMERA CIRCUITS

This chapter describes the theory of operation for typical camcorder camera circuits. In addition to camera basics, the circuits described here include power distribution, deflection, and signal processing.

By studying the circuits found in this chapter, you should have no difficulty in understanding the schematic and block diagrams of similar camcorders. This understanding is essential for logical troubleshooting and repair, no matter what type of electronic equipment is involved.

No attempt has been made to duplicate the full schematics for all circuits. Such schematics are found in the service literature for the particular camcorder. Instead of a full schematic, the circuit descriptions are supplemented with partial schematics and block diagrams that show such important areas as signal flow paths, input/output, adjustment controls, test points, and power source connections. These are the areas most important in service and troubleshooting. By reducing the schematics to these areas, you will find the circuit easier to understand, and you will be able to relate circuit operation to the corresponding circuit of the camcorder you are servicing.

4-1 CAMERA BASICS

Figure 4-1 is a block diagram of the basic camera circuits. Although the camcorder shown in Fig. 4-1 uses a Saticon tube, the same circuits are found in camcorders with Newvicon tubes. Likewise, some of the circuits are the

Sec. 4-1   Camera Basics   69

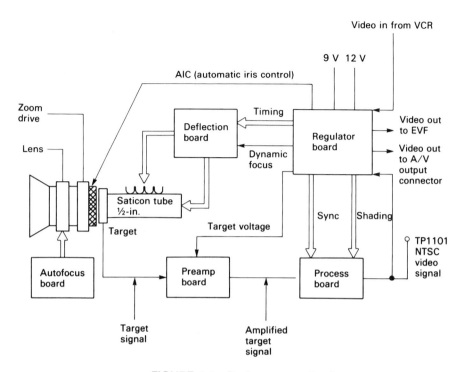

FIGURE 4-1 Basic camera circuits

same for camcorders with MOS pickups. We discuss any circuit differences between tube and MOS pickups in Sec. 4-6.

The particular camcorder of Fig. 4-1 has five circuit boards in the camera section. Most of these boards are located in the camcorder handle. (This arrangement is typical for many video cameras.)

The *deflection board* includes circuits that bias the pickup tube, as well as provide power for both the horizontal and vertical deflection of the tube beam (through the horizontal and vertical windings of the deflection yoke).

The deflection circuits receive timing and synchronization (sync) signals from the master sync generator on the *regulator board*. Various types of power supply regulators, B+ switching circuits, and video switching circuits are also located on the regulator board.

The signal (or carrier) from the pickup tube is removed by the target contact (the ring around the front of the tube) and is applied to the *preamp board*. The target signal is amplified by circuits on the preamp board. The amplified target signal is then routed to the *process board*. The majority of the signal processing for the target signal is done on the process board.

One of the main functions of the process board is to convert the target signal into an NTSC video signal. Note that the camcorder of Fig. 4-1 has a test point (TP1101) to monitor the NTSC video signal. Virtually all camcorders have a similar test point (although the test point may not have the same reference designation or be located on a process board). The test point is quite important in troubleshooting and is comparable to the "looker point" at the output of the tuner in a TV set.

The NTSC signal from the process board is passed back to video switching circuits on the regulator board. This video is applied to the EVF and is also available at the audio/video output connector (and can be applied to a TV or VCR through an RF adapter, as discussed in Chapter 2).

The optics/lens assembly of the camcorder uses an infrared (IR) autofocus system. Circuits on the *autofocus board* (usually mounted on the lens assembly) generate an IR signal that is transmitted through the lens to the object being photographed. The reflected IR signal is passed to sensors (photodiodes) that operate the autofocus motor. When autofocus is selected, the motor drives the lens focus ring as necessary to produce correct focus. The lens can also be focused by a power zoom drive. As with most video cameras, there are two zoom buttons (on or near the handle grip) or a single rocker-type button with two positions. One position is for telephoto; the other is for wide angle.

The lens assembly also has a variable opening aperture called the iris (or iris assembly). This iris is controlled by signals from the regulator board. The variable iris limits the amount of light that reaches the pickup tube target surface. This makes it possible to have a very sensitive target material, but still protect the target from excessive light. When light is strong, the iris closes to limit light hitting the target. The opposite occurs when the light is weak.

4-1.1 Camera Optics and Lens Assembly

It is assumed that you have read Chapter 1 and that you are familiar with the basics of light, color temperature, and the striped filter used in the optics of camcorders with pickup tubes. (MOS pickups are discussed in Sec. 4-6.) So we will not dwell on these subjects here. Also, in most camcorders, the optics or lens must be replaced as a complete assembly (exclusive of the iris, zoom motor, focus motor, lens hood, pickup tube retainer or rear block, switches, and circuit boards). So we will not go into details of lens construction.

However, it may be of some practical help to understand the optics between the incoming light (from the scene being photographed) and the pickup tube. So let us go through this before we get to camera circuit details. Figure 4-2 shows the sequence of light through the optics or lens assembly to the pickup tube.

Sec. 4-1 Camera Basics   71

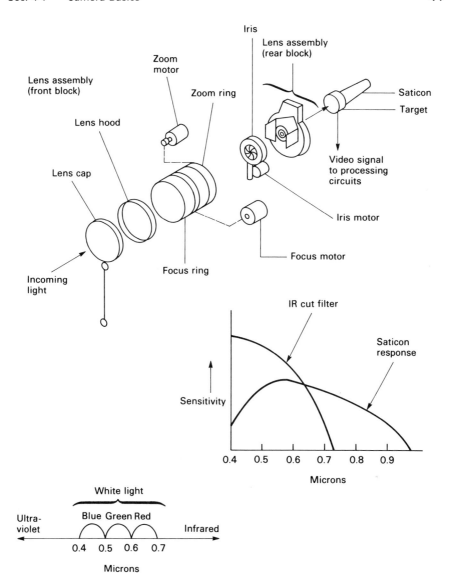

FIGURE 4-2 Sequence of light through optics to pickup tube

In addition to the color temperature discussed in Chapter 1, another important consideration for light is the wavelength. Light is made up of electromagnetic waves that have very short wavelength. As shown in Fig. 4-2, electromagnetic waves with a length of about 0.4 to 0.5 micron (micron = 0.000001 meter) are seen by the human eye as blue light. Light with wavelengths of about 0.5 to 0.6 micron appear as green light. Wavelengths of about 0.6 to 0.7 micron are seen as red light. White light is produced when all these wavelengths are combined in the proper ratio.

Waves that are shorter than about 0.4 micron (ultraviolet) or longer than about 0.7 micron (infrared) are generally not detected by the human eye. As a result, the pickup target must be sensitive only to light waves in the range from about 0.4 to 0.7 micron, if the correct color video signal is to be generated. Unlike the human eye, the target material or most pickup tubes is sensitive to IR light and produces erroneous signals when subjected to IR radiations. To prevent this condition, all IR is filtered out by an IR cut-filter between the lens and target. Figure 4-2 shows typical IR cut-filter response.

The light reflected from the subject is passed to the lens assembly through a lens hood. The lens assembly has two adjustable rings. Adjusting the first (or focus) ring alters the position of the lenses so that the image being viewed is properly focused on the target surface of the pickup tube. From a troubleshooting standpoint, remember that this has nothing to do with focus of the beam (on the other side of the target) within the tube. If the focus ring is improperly adjusted, the image will be blurred. Also keep in mind that the focus ring can be adjusted manually or automatically and the focus motor can be replaced as a separate component (on most camcorders).

The second adjustable ring on the lens provides a variable zoom control of the lens assembly. Adjusting the zoom ring controls a 6 to 1 magnifying lens system. As a result, the image can be changed in perceptive distance by a ratio of 6 to 1. As in the case of focus, the zoom ring can be adjusted manually or automatically, and the zoom motor can be replaced as a separate component.

Also included in the lens assembly is the iris that controls that amount of light passing through to the pickup target. In most camcorders, the iris is replaced as an assembly or block (including the iris motor) separate from the lens assembly. Generally, it is necessary to remove the rear block shown in Fig. 4-2 to replace the iris. As usual, never try to remove or replace any part of the lens assembly or pickup tube without consulting the service literature. Camera optics are all similar, but *not identical*!

After passing through the lenses, iris, and IR filter, incoming light information is converted into an electrical signal by the pickup tube. As discussed in Chapter 1, the tube is much like a conventional cathode ray tube, with the exception that the electron beam strikes a photosensitive material rather than a luminescent phosphor (as is the case with the cathode ray tube).

As the beam is scanned across the target surface, a current is generated proportionally to the light falling upon that portion of the target. The target current is supplied (via the target lead) to the preamplifier circuit where a video signal is generated.

Sec. 4-1   Camera Basics   73

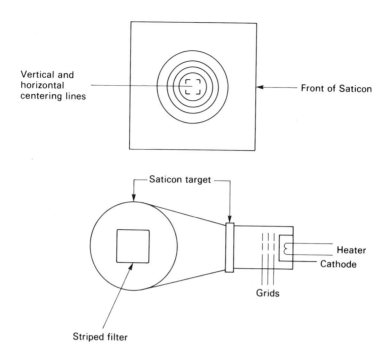

FIGURE 4-3   Details of Saticon tube and striped filter

As shown in Fig. 4-3, the target surface of the pickup tube is a rectangular-shaped area in the center of the tube. Note that it is the target surface that is sensitive to the incoming light which generates the current. If you are curious, you can see the target surface by removing the pickup tube and looking directly at the tube front. In the case of the Saticon tube shown in Fig. 4-3, you can see four lines pointing in from both sides as well as the top and bottom of the tube. These are the vertical and horizontal centering lines used during alignment to position the vertical and horizontal scan properly across the target surface. In most cases, it is necessary to align the tube mechanically and then adjust electrically for proper centering. Always check the service literature.

The author suggests that you not become too curious about the pickup tube or lens assembly. Unless required for service, do not pull the tube, or disassemble the lens. If you must handle the tube, take great care to avoid touching or contaminating the front surface.

With any type of pickup, fingerprints or small specks of dust block the incoming light from reaching the target surface. This can create distortions and spots in the video. If dust is present on the front of the pickup tube, wipe the dust with a very soft lint-free cloth. Always handle the pickup tube with extreme care.

4-2 CAMERA POWER DISTRIBUTION CIRCUITS

Figure 4-4 shows the power distribution for the camera section of a typical camcorder. We discuss power distribution for the entire camcorder in Chapter 7.

Power (+12 V) for the camera section of the camcorder is taken from power on/off relay RL901 (Chapter 7) and is applied to the camera regulator board. As shown in Fig. 4-4, most of the camera power distribution components are on the regulator board.

Power is first applied to 8-V regulator IC1604, which is turned on and off by a control signal from system-control IC901 (Chapter 7). When the camera/VCR-control signal from IC901 is high (camera), IC1604 is turned on through Q1507/Q1510 and supplies regulated 8-V power to the various components.

The +12-V power from RL901 is also applied to standby/on switch S1602. When S1602 is set to on (normal), +12-V power is applied to the input of B+ switch Q1503 and to other components on the regulator board.

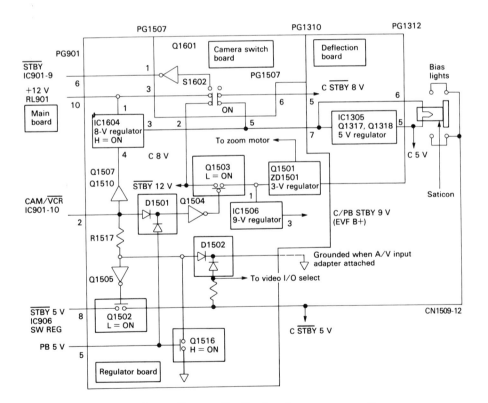

FIGURE 4-4 Power distribution for camera section

Sec. 4-2   Camera Power Distribution Circuits

Q1503 is controlled by signals applied through D1501 and Q1504. Diode D1501 is a dual-input diode with two anodes and a common cathode.

One input is from the camera/VCR control line. In the camera mode, the line high is applied through D1501 to Q1504, where the high is inverted to a low. The low is applied to Q1503, turning Q1503 on. This applies +12 V to the input of IC1506, a 9-V regulator, and to the input of 3.6-V regulator, Q1501/ZD1501. The 9-V power is applied to the EVF, while the 3.6-V power is used by the zoom motor.

During playback, the PB5-V line goes high. The high is applied through the other diode in D1501 and inverted to a low by Q1504, turning Q1503 on. This applies 9-V power to the EVF during playback.

In the CAMERA mode, the camera/VCR line high is applied to Q1505 where the high is inverted to a low, turning Q1502 on. This transfers standby 5-V power from IC906 in the main circuit board (Chapter 7) to the camera/standby 5-V distribution line. As discussed in Chapter 7, standby 5-V power from IC906 is turned off when S1602 is set to standby. The standby power can also be interrupted to the distribution line by signals applied through Q1502 or D1502.

During VCR playback, Q1516 is turned on by the playback 5-V signal (PB 5V). This grounds the input to Q1505 and keeps Q1502 open, removing the 5-V standby power from the camera circuits.

When the audio/video adapter is connected to the EVF connector in place of the EVF (Chapter 6), the diodes in D1502 are grounded through the adapter. This grounds the input to Q1505 and opens Q1502, to remove the standby power.

The regulated 8-V power from IC1604 is also applied to a 5-V regulator consisting of IC1305, Q1317, and Q1318. The 8-V and 5-V lines are connected to the filaments of the pickup tube. (The 3-V difference powers the tube filaments.) The regulated 5-V power is also applied to various components on the deflection and process boards.

4-2.1 Camera Power Circuit Troubleshooting

If problems are suspected in the camera section of the camcorder, the related power distribution circuits should be checked first (as is the case with most circuits). Keep in mind that the power circuits are interrelated with the system-control circuits (Chapter 7).

As an example, when S1602 is set to standby, 12-V battery power is inverted by Q1601 and is applied to the system-control microprocessor IC901 as a standby (or power-save) signal. In turn, IC901 cuts off IC906, removing the standby power to Q1502.

First, check for 8V at the output of IC1604 (at IC1604-3 or at any point on the 8-V line) and for 12 V at the input of IC1604 (at IC1604-1).

If the 12-V power is missing, suspect the battery or system control

(Chapter 7). If the 8-V power is missing, with the 12-V input present, check for a high on the camera/VCR line and at pin 4 of IC1604.

If the high is missing, suspect system control (Chapter 7). If the camera/VCR line is high, but pin 4 of IC1604 is not, suspect Q1507/Q1510.

Next, check for 9-V power to the EVF at pin 3 of IC1506 and for 12-V power at the input of IC1506 (at IC1506-1). If the 12-V power is missing (with S1602 set to on and a good battery), suspect Q1503, Q1504, or D1501. If the 9-V power is missing, with 12-V present, suspect IC1506.

Next, check for 3.6-V power to the zoom motor through Q1501/ZD1501. If the 3.6-V power is missing, with 12-V power available from Q1503, suspect Q1501/ZD1501.

Next, check for regulated 5-V power to the pickup tube filaments and to various components on the deflection and process boards. If the regulated 5 V is missing, with 8 V present, suspect IC1305 and Q1317/Q1318.

Finally, check for standby 5-V power at both input and output of Q1502. If there is no 5 V at the input, suspect the system-control circuits (Chapter 7). If there is no 5 V at the output of Q1502, with 5 V at the input and the camera/VCR line high, suspect Q1502, S1505, or R1517. Of course, if you are in playback, the PB 5-V signal grounds the input to Q1505 and keeps Q1502 open. Likewise, if the audio/video adapter is used (in place of the EVF), Q1505 is grounded and Q1502 is open.

4-2.2 Bias Lights

As shown in Fig. 4-5, the bias lights are four LEDs mounted around the pickup tube near the target surface. The bias lights receive their power from the camera power distribution circuits (the standby 5-V power, Fig. 4-4).

The bias LEDs maintain a constant low-level light on the pickup tube, except during standby and playback or when the audio/video adapter is used. The bias lights prevent the iris from opening fully, even in the dark. This

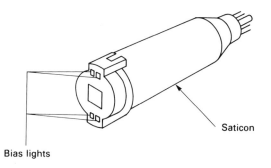

FIGURE 4-5 Typical bias light mounting configuration

Sec. 4-3   Camera Deflection Circuits   77

prevents the pickup tube from suddenly being overdriven when saturated with intense light (say, when the lens cap is pulled off in bright sunlight).

From a troubleshooting standpoint, there is usually no problem with the bias lights (although it is possible that one or more LEDs can burn out). Note that the bias light is adjustable on some camcorders. Typically, the adjustment consists of monitoring the prevideo signal (Sec. 4-4) and adjusting the light intensity for a given signal amplitude.

4-3 CAMERA DEFLECTION CIRCUITS

Figure 4-6 shows the deflection circuits for the camera section of a typical camcorder. Readers familiar with video cameras will recognize most of the circuits.

Sync generator IC1106 produces a variety of crystal-controlled sync signals used by the deflection circuits (and the signal-processing circuits, Sec. 4-4). The three major signals produced by IC1106 are the wide horizontal drive (WHD), the vertical pulse (VP), and the clamp pulse 2 (CP2). These signals ensure that all functions of the deflection and signal-processing circuits are synchronized as to time.

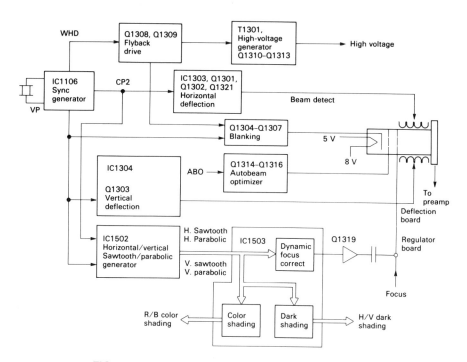

FIGURE 4-6 Deflection circuits for camera section

The WHD signal is applied to the flyback drive circuit Q1308/Q1309. In turn, the flyback drive circuit powers the high-voltage circuit T1301, Q1310-Q1313. Output voltages from the high-voltage circuit are used to bias the grids of the pickup tube. An additional output from the flyback drive is to the blanking amplifier Q1304-Q1307.

The VP signal is applied to the blanking amplifier Q1304-Q1307; to the vertical deflection circuit IC1304, Q1303; and to the horizontal/vertical sawtooth and parabolic signal generator IC1502. The blanking amplifier Q1304-Q1307 turns off the cathode emission of the pickup tube during the vertical and horizontal blanking periods of the scan.

A beam-detect system is associated with the blanking circuit. The beam detect is used by the automatic iris control (AIC) circuits (Sec. 4-4) as a safety feature. The AIC circuits open the iris lens only after the pickup tube emission (cathode current) is detected. If a problem exists with the pickup tube (no beam), the AIC circuits prevent the iris from opening.

The CP2 signal is applied to IC1502 and to the horizontal deflection circuits IC1303, Q1301, Q1302, Q1321.

The autobeam optimizer (ABO) circuit receives an input from the signal process circuits (Sec. 4-4) and controls the pickup tube beam so that the beam always matches the object.

Those familiar with tube-type color cameras will know that some type of raster and shading correction must be used to get a completely white raster on a monitor when the camera is viewing a white scene. These correction functions are provided by IC1502 and IC1503 in the camcorder of Fig. 4-6.

IC1503 uses four signals from IC1502 to produce the necessary correction signals (color shading, dark shading, and dynamic focus). The four signals from IC502 include horizontal sawtooth, horizontal parabola, vertical sawtooth, and vertical parabola.

The dynamic focus signal from IC1503 is amplified by Q1319 and is mixed with the d-c focus signal to provide for optimum focus as the beam scans the target area.

The red/blue color-shading correction signals are applied to the signal-processing circuits (Sec. 4-4) and compensate for the output variations from the target surface under different light conditions.

The horizontal/vertical dark-shading correction signals are also applied to the signal-processing circuits (Sec. 4-4) and compensate for variations in luminance sensitivity of the target area.

4-3.1 Sync Generator

Figure 4-7 shows typical sync generator circuits. Figure 4-8 shows typical timing pulses.

IC1106 is controlled by crystal X1101 and is adjusted by trimmer CT1101 to oscillate at a 14.31818-MHz reference frequency. This reference is divided

Sec. 4-3   Camera Deflection Circuits

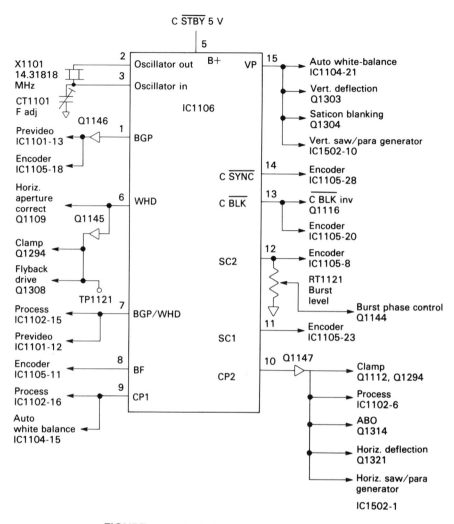

FIGURE 4-7 Typical sync generator circuits

down to produce 3.58-MHz (subbarrier), 15.734-kHz (horizontal sync), and 60-Hz (vertical sync) signals. A composite sync signal is generated from the horizontal and vertical sync signals.

4-3.2 Horizontal Deflection Circuits

Figure 4-9 shows typical horizontal deflection circuits.

The clamp pulse CP2 from the sync generator is applied to the horizontal deflection driver Q1302 through buffer Q1321 and inverter Q1301. Flutter control RT1330 sets the level of the CP2 signal. During the low to high

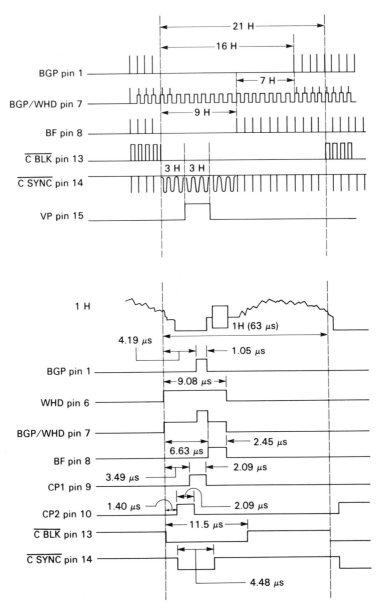

FIGURE 4-8 Typical timing pulses

(blanking) transition of the inverted CP2 pulse, Q1302 turns off, allowing direct current to flow from the 5-V source through the horizontal size control RT1317, L1301, and C1308 to the horizontal deflection yoke. At the same time, the horizontal drive pulse from Q1302 is supplied to IC1303 at pins 5 and 6, making the output of IC1303 (at pin 7) low.

Sec. 4-3　Camera Deflection Circuits

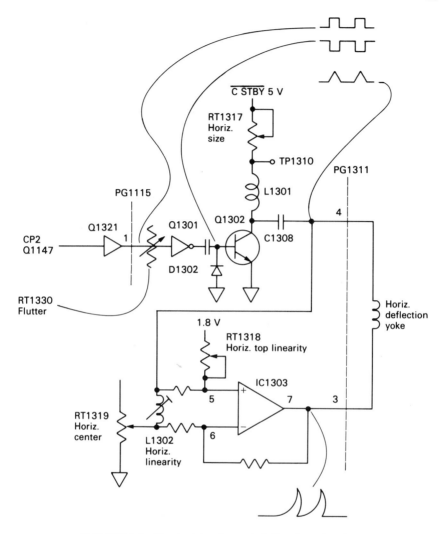

FIGURE 4-9 Typical horizontal deflection circuits

During the CP2 pulse high duration (scanning interval), Q1302 turns on, and a sawtooth output from IC1303-7 flows to ground (through the horizontal deflection yoke, C1308 and Q1302). This produces the usual beam scan, similar to the horizontal scan in a TV picture tube.

The horizontal controls are also similar to those of a TV picture tube. Both horizontal top linearity control RT1318 and horizontal linearity control L1302 affect the sawtooth current output from IC1303 and, thus, determine linearity of the horizontal sweep. Horizontal centering control RT1319 sets the fixed reference voltage applied to IC1303 and, thus, sets the horizontal position of the beam on the target surface. The horizontal size control

RT1317 controls the amount of direct current through the horizontal deflection yoke and, thus, determines the size (or width) of the horizontal scan.

Note that the horizontal circuits receive power from the camera standby 5-V distribution. This eliminates the horizontal sweep during standby and playback and when the A/V adapter is used (Sec. 4-2).

4-3.3 Vertical Deflection Circuits

Figure 4-10 shows typical vertical deflection circuits.

The vertical pulse VP from the sync generator is applied to the vertical deflection circuits through switch Q1303. The VP pulse causes Q1303 to turn on and off at a 60-Hz rate.

When the VP pulse is low, Q1303 turns off, causing the voltage at pin 2 of IC1304 to increase as C1313 is charged. As the voltage at IC1304-2 increases, the output of IC1304-1 decreases.

When the VP pulse is high, Q1303 turns on, allowing C1313 to discharge. The charge and discharge of C1313 produces a sawtooth sweep at the input

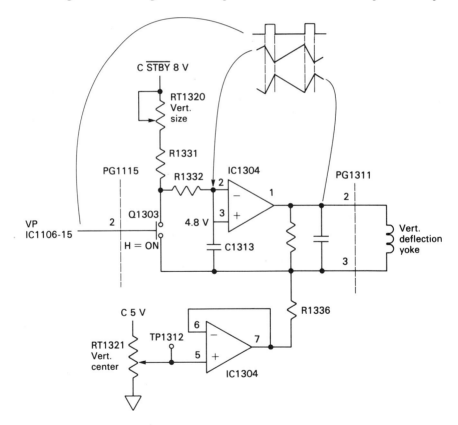

FIGURE 4-10 Typical vertical deflection circuits

Sec. 4-3   Camera Deflection Circuits   83

and output of IC1304. The output sweep is applied through the vertical deflection coil and produces vertical deflection similar to that of a TV picture tube.

The vertical controls are also similar to those of a TV picture tube. Vertical size control R1320 determines the charge time of C1313 and, thus, determines the size of the vertical sweep. The vertical center control R1321 sets the fixed reference voltage applied to the vertical deflection yoke (through IC1304) and, thus, sets the vertical position of the beam on the target surface.

4-3.4 Horizontal/Vertical Sawtooth and Parabolic Signal Generator

Figure 4-11 shows typical horizontal/vertical sawtooth and parabolic signal generator circuits.

This circuit generates the horizontal and vertical sawtooth and parabolic signals for generating the correction signal and for generating the window gate pulse applied to the iris control in the signal processing circuits (Sec. 4-4).

The horizontal sawtooth is produced by applying CP2 to SW1 in IC1502. When CP2 is low, SW1 turns off, causing the charging current to flow

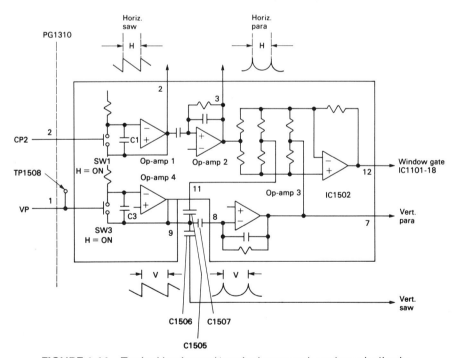

FIGURE 4-11 Typical horizontal/vertical sawtooth and parabolic signal generator circuits

through ramp capacitor C1. The voltage at the inverting input of op-amp 1 increases as C1 charges, so the output of op-amp 1 decreases.

When CP2 is high, SW1 turns on, causing C1 to discharge through SW1. The repeated charging and discharging of C1 produces the horizontal sawtooth signal, which is output at pin 2 of IC1502.

The horizontal signal is also applied to the *horizontal parabolic signal generator* in IC1502. Op-amp 2 and C2 produce the horizontal parabolic signal, which is output at pin 3 of IC1502.

The *vertical sawtooth* is produced by applying VP to SW3 in IC1502. Op-amp 4 and SW3 operate in essentially the same way as op-amp 1 and SW1 to produce the vertical sawtooth signal, which is output at pin 9 of IC1502.

The vertical signal is also applied to the *vertical parabolic signal generator* in IC1502 through C1507. Op-amp 3 and C4 produce the vertical parabolic signal, which is output at pin 7 of IC1502.

Both the horizontal and vertical parabolic signals, as well as the vertical sawtooth signal, are applied to the *window gate pulse generator* in IC1502.

The three inputs are applied to a differential amp that detects the center of the vertical and horizontal periods of the video signal and produces a corresponding window gate pulse. The gate pulse is output from pin 12 of IC1502 and is applied to the gate circuit in the iris control of the signal-processing circuits (Sec. 4-4).

4-3.5 Dynamic Focus Correction Circuit

Figure 4-12 shows typical dynamic focus correction circuits.

Generally, the level of focus voltage that produces the best focus of the pickup tube beam in the center part of the tube is different from that which produces the best focus at the edges of the pickup tube. As a result, the modulation depth for the center of the tube is different from the depth at the edges. When the camera is directed at an evenly illuminated white object, the red and blue signals (modulated at 3.58 MHz) do not have a uniform level. As a result, color shading appears (typically a greenish color shading).

The dynamic focus circuits correct this condition by varying the d-c focus voltage (also applied to grid 4 of the pickup tube) as required to offset any unevenness in modulation.

The dynamic focus circuits use the four signals from IC1502 (H/V sawtooth and H/V parabolic) to produce the correction signals. The four signals from IC1502 are applied to a differential amp within IC1503 through corresponding adjustment controls (RT1505, RT1506, RT1507, and RT1508). The signals overlap and produce a synthesized differential signal, which is amplified by an op-amp within IC1503 and inverted by Q1319. The correction signal is combined with the d-c focus voltage to produce an even depth of modulation across the scanning area.

Sec. 4-3 Camera Deflection Circuits

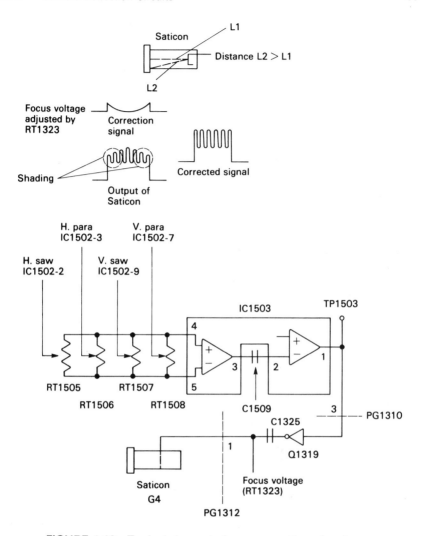

FIGURE 4-12 Typical dynamic focus correction circuits

4-3.6 Pickup Tube Blanking and Beam-Detection Circuit

Figure 4-13 shows typical blanking and beam-detection circuits.

This circuit generates blanking pulses and applies the pulses to the pickup tube cathode, thus eliminating scanning retrace lines.

The WHD pulses from sync generator IC1106 are converted to flyback pulses by Q1308, Q1309, and D1303 and are applied to Q1306. The VP pulses are inverted by Q1304 and are applied to Q1305.

During the blanking interval, the inverted VP signal turns Q1305 on,

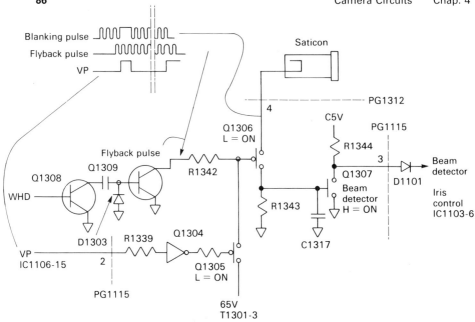

FIGURE 4-13 Typical blanking and beam-detection circuits

applying 65 V from high-voltage transformer T1301 (Sec. 4-3.8) to Q1306. This turns Q1306 off and blanks the pickup tube. The pickup tube is also blanked when the flyback pulses are high.

When both the VP and flyback pulses are low, the blanking function is removed, Q1306 is on, and scanning occurs.

When the pickup tube filaments are warming up, or when power is low, the pickup tube beam is absent or weak. This causes the iris to open. If you take pictures of high-luminance objects under these conditions, the pickup tube can be burned out. The beam-detection circuit prevents such a disaster.

The beam is on only when Q1306 is on. Under these conditions, C1317 is charged, Q1307 is turned on, and the junction of R1344 and D1101 is at ground. This ground or low keeps the iris open (Sec. 4-4).

When Q1306 is open (during blanking), or when there is very little cathode current (during warmup), C1317 cannot charge, Q1307 is turned off, and a high is applied to the iris through R1344 and D1101. This high shuts the iris and prevents light from reaching the pickup tube.

4-3.7 ABO (Autobeam Optimizer) Control Circuits

Figure 4-14 shows typical ABO control circuits.

The ABO circuit controls the beam of the pickup tube so that the beam always matches the object light. Output from the ABO circuit is

Sec. 4-3   Camera Deflection Circuits   87

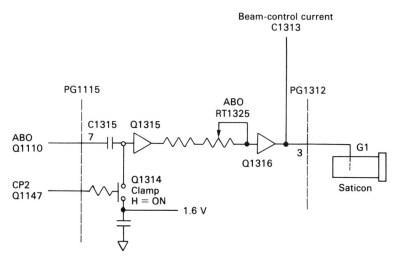

FIGURE 4-14 Typical ABO control circuits

mixed with beam control current from Q1313 in the high-voltage circuit (Sec. 4-3.8) and adds to or subtracts from the potential applied to grid G1 of the pickup tube.

Input to the ABO circuit is an ABO signal from the signal-processing circuits (Sec. 4-4). Variations in light cause the ABO signal to vary. In turn, the ABO circuits vary the pickup tube beam to maintain the beam at an optimum level.

The base of Q1316 is maintained at a fixed 2.3 V, while the base of Q1315 is clamped to 1.6 V by Q1314 when a CP2 pulse is applied. Since the base of Q1316 is fixed, variations in the ABO input signals cause corresponding variations in the collector current of Q1316 and, thus, vary the pickup tube beam. The amount of control produced by the ABO circuit is set by RT1325.

4-3.8 High-Voltage Power Supply

Figure 4-15 shows typical high-voltage power supply circuits for the pickup tube.

This circuit provides voltages for the pickup tube grids and target. The circuit also provides a fixed 65-V potential and a flyback pulse for the blanking and beam-detection circuit (Sec. 4-3.6).

Inputs to the ABO circuit include the WHD pulse from the sync generator IC1106, the ABO control signal (Sec. 4-3.7), and the dynamic focus signal (Sec. 4-3.5).

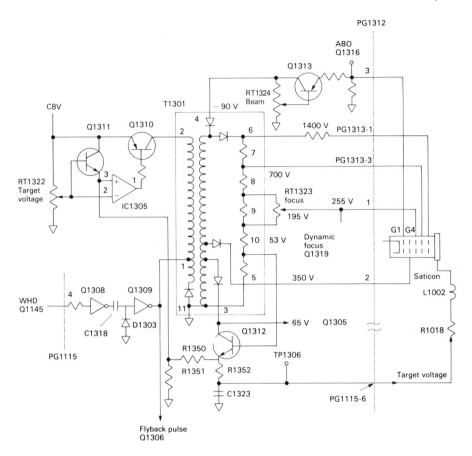

FIGURE 4-15 Typical high-voltage power supply circuits

The 8-V source is applied to the start-up circuit composed of IC1305, Q1310, and Q1311. T1301 contains a step-up transformer and several rectifiers to produce the d-c voltages required to power the pickup tube.

RT1322 sets the target voltage by adjusting the voltage on the base of Q1311 and on the inverting input of IC1305. Note that the target voltage is applied to the pickup tube through R1018 and L1002 on the preamp board (Sec. 4-4).

RT1324 sets the pickup tube beam intensity by adjusting the voltage on the base of RT1324, thus controlling the voltage on grid G1 of the pickup tube. Note that grid G1 also receives the ABO output signal from Q1316.

RT1323 sets the pickup tube focus by adjusting the voltage on grid G4. Note that grid G4 also receives the dynamic focus signal from Q1319.

Sec. 4-3 Camera Deflection Circuits 89

4-3.9 Camera Deflection Circuit Troubleshooting

Problems in the camera deflection circuits usually show up as picture problems in the EVF and/or monitor. For example, typical picture problems include no horizontal deflection, no vertical deflection, poor focus, erratic blanking, or no picture (indicating that the target carrier is totally dead).

Before you condemn the deflection circuits, check operation on both a TV and the EVF. If the deflection appears normal on the TV, the problem is likely in the EVF circuits rather than the deflection circuits. If both the TV and EVF show the same abnormality, suspect the deflection circuits.

Next, check that all power sources are correct and that signals from the sync generator IC1101 are present. For example, if there is no CP2 signal, there can be no horizontal deflection. If there is no VP signal, there can be no vertical deflection, and so on.

If the picture is totally absent, check all the voltages from the high-voltage power supply (Fig. 4-15). Then try adjusting the high-voltage circuits. For example, if the target voltage is absent or abnormal, adjust RT1322. If focus is poor, adjust RT1323. If the G1 grid voltage is absent or abnormal, adjust RT1324. If all the voltages are correct, try replacing the pickup tube.

If focus appears abnormal, and cannot be corrected by adjustment of R1323 on the high-voltage circuits, suspect the dynamic focus correction circuit (Fig. 4-12). First make certain that the four drive signals from the H/V sawtooth and parabolic signal-generating circuit (Fig. 4-11) are present and normal. Then try correcting focus problems by adjustment of RT1505 through RT1508. There should be a measurable change in pickup tube G4 voltage as any of the dynamic focus adjustments are made.

If there is no vertical deflection, check for drive signals to the vertical coil of the deflection yoke. Figure 4-10 shows the approximate waveforms. If the drive signals are present, suspect the yoke. If the drive signals are absent or abnormal, try adjusting RT1320 (for vertical size) and RT1321 (for vertical centering).

If there is no horizontal deflection, check for drive signals to the horizontal coil of the deflection yoke. Figure 4-9 shows the approximate waveforms. If the drive signals are present, suspect the yoke. If the drive signals are absent or abnormal, try adjusting RT1317 (for horizontal size), RT1319 (for horizontal centering), and L1302/RT1318 (for horizontal linearity).

If you get "comet tails" or streaks in high-luminance areas, or no picture in low-light areas, suspect the ABO circuits (Fig. 4-14). Try adjusting R1325. There should be a measurable change in pickup tube G1 voltage as R1325 is adjusted.

If blanking appears to be abnormal, check for proper blanking pulses at the pickup tube cathode. Figure 4-13 shows typical blanking pulses. Keep in mind that if there is a problem in the blanking circuits, the beam defect can hold the iris closed. For example, if Q1306 is on (closed) but Q1307 is off (open), a high is applied to the iris and keeps the iris closed.

4-4 CAMERA SIGNAL-PROCESSING CIRCUITS (WITH PICKUP TUBE)

Figure 4-16 shows the signal-processing circuits for the camera section of camcorders with pickup tubes. Readers familiar with video cameras will recognize many of the circuits.

The signal from the target surface is amplified by a low-noise *preamp system*. The output signal from the preamp is applied to the *prevideo processing/tracking signal generator* within IC1101. A *dark-shading* correction signal from the regulator is applied to the prevideo processor. The target signal from the prevideo processing circuit is applied to a gain-up amplifier.

The output signal from the tracking signal generator circuit is applied to the *tracking/shading correction* circuit within IC1104, along with the *color-shading* correction signal from the regulator. These correction signals are used to optimize gain of the *red and blue process amplifier* in IC1102 to compensate for target surface output variations under different light levels.

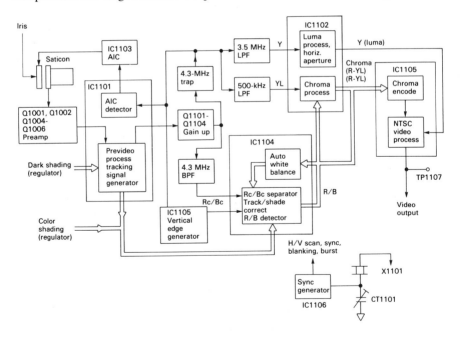

FIGURE 4-16 Signal-processing circuits for camera section

The carrier output signal from the gain-up amplifier is applied to both a 4.3-MHz trap and a 4.3-MHz bandpass filter (BPF). The 4.3-MHz BPF passes the red and blue carrier signals (Rc/Bc) to the input of IC1104. The 4.3-MHz trap output is applied to the AIC detector in IC1101, to the vertical edge generator in IC1105, and to the luminance-processing circuits in IC1102.

The 4.3-MHz signal to the AIC detector represents the target signal level and, thus, varies with light. The AIC detector output is applied to the iris through the AIC amplifier in IC1103. The iris is opened and closed as necessary to compensate for variations in light reaching the target (light increases close the iris, and vice versa). This keeps the target light constant.

The 4.3-MHz signal to IC1102 is applied through a 3.5-MHz LPF and a 500-kHz LPF to produce Y- and YL-signals, respectively. The signal component below 500 kHz is the YL-signal, which can be considered to be the *green element* of the picture. The signal component at 3.5 MHz and lower is referred to as the Y or *luminance information*.

Within IC1102, the red and blue chroma signals from IC1104 are combined with the YL-signals to produce R-YL- and B-YL-signals. These output signals are passed to the *chroma encoder* portion of IC1105 and to the *automatic white-balance* circuits of IC1104.

The Y- or luminance signal is processed by compensation circuits within IC1102 and is applied to the NTSC *video process amplifier* within IC1105. The chroma and luminance signals, along with various sync signals, are combined in IC1105 to produce a standard NTSC video signal. In the camcorder of Fig. 4-16, the composite NTSC signal can be monitored at TP1107.

The *vertical-edge generator* produces a correction signal that is applied to the red and blue carrier amplifier within IC1104. The correction signal shifts amplifier gain accordingly when a vertical edge is detected. This minimizes *vertical-edge distortion* occurring on certain types of scenes (horizontal stripes, where there is a sharp, white-to-black or black-to-white vertical transition).

4-4.1 Preamplifier Circuit

Figure 4-17 shows typical preamplifier circuits.

This circuit converts weak signal current from the pickup tube to a voltage signal and then amplifies the voltage signal into a prevideo signal of appropriate level. The circuit has one input from the pickup tube and produces one output at the emitter of Q1001 to the prevideo processor circuits (Sec. 4-4.2). The circuit also provides for application of the target voltage from the high-voltage power supply to the pickup tube target.

The preamp has a low output impedance and uses negative feedback in an amplifier with a low-noise JFET input (to improve the signal-to-noise ratio). CT1001 (called the *smear* or *streaking* adjustment) varies the low-

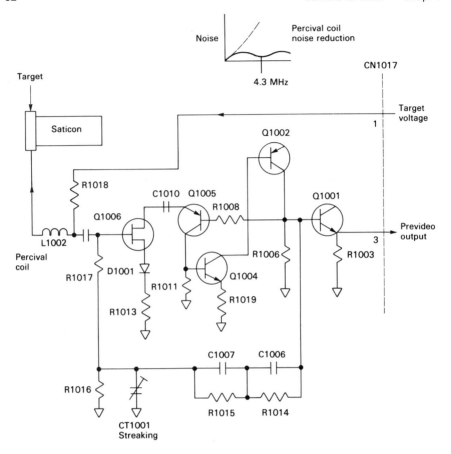

FIGURE 4-17 Typical preamplifier circuits

frequency (400- to 500-kHz) portion of the negative feedback signal to maintain a flat frequency response.

A *Percival compensation circuit* is used at the input to prevent S/N deterioration due to impedance mismatch. The Percival coil L1002 resonates at about 4.3 MHz and separates the pickup tube output capacitance from the preamp input capacitance.

4-4.2 Prevideo Processing Circuit

Figure 4-18 shows typical prevideo processing circuits.

The prevideo processing circuits apply various corrections to the video signal from the preamp. In addition to the prevideo input, the circuit receives horizontal/vertical sawtooth and parabolic signals (Sec. 4-3.4), Y-signals from the AIC/AGC circuits (Sec. 4-4.3), and various input pulses. The main output is to the ABO control circuits (Sec. 4-3.7). The processing circuits

Sec. 4-4 Camera Signal-Processing Circuits (with Pickup Tube)   93

also produce a high/low-luma/chroma clip signal to the chroma encoder (Sec. 4-4.9) and automatic white-balance circuits (Sec. 4-4.11).

The prevideo signal is applied through low-light gain control RT1103 to IC1101. RT1103 varies the level of the prevideo signal to match the input at IC1101. The CP2 pulse turns Q1112 on and eliminates spiked pulses during the blanking interval.

Unbalanced characteristics in the pickup tube, and unbalance bias light quantities, cause the black level of the prevideo signal to fluctuate. This is called *dark current dispersion*. A dark-shading correction signal is applied to prevent shading caused by processing the unmodified prevideo signals. The correction signal is generated by adjusting the amplitude of the H. SAW and H. PARA signals (Sec. 4-3.4) through RT1501, RT1502, and IC1503. The correction signal is applied through pin 2 of IC1101 to the prevideo signal. The signal is then applied to a feedback clamp within IC1101 for vertical dark-shading correction.

The feedback clamp circuit also receives a vertical dark-shading correction signal from the V. SAW and V. PARA signals (Sec. 4-3.4) through RT1503, RT1504, and IC1503. The correction signal is superimposed on an AGC offset voltage at Q1111. The AGG offset is adjusted by RT1101. The vertical correction signal acts as a reference for the feedback clamp.

When a BGP pulse is input through pin 13 of IC1101 to the feedback clamp, the black level of the prevideo signal is compared to the reference (at pin 8 of IC1101), and a differential voltage is generated. The differential

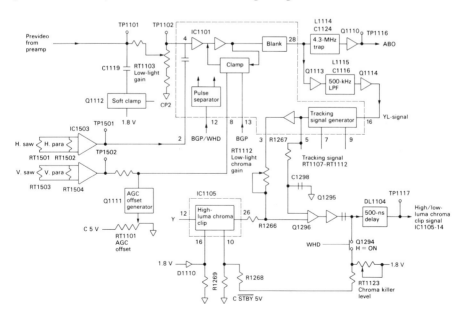

FIGURE 4-18 Typical prevideo processing circuits

voltage is applied to the prevideo signal at a point in the signal path between the two buffers. The BGP pulse is taken from the pulse separator within IC1101. The pulse separator separates the superimposed BGP/WHD pulses applied to pin 12 of IC1101 and applies the pulses to circuits within IC1101.

The prevideo signal is then applied to a blanking circuit within IC1101. When a WHD pulse is present (blanking interval), the blanking circuit removes noise. Output from the blanking circuit is applied to an AGC circuit within IC1101 (Sec. 4-4.3) and to the ABO circuit through a 4.3-MHz trap (L1114/C1124) and buffer Q1110. (The ABO circuit varies the pickup tube beam to match the scene, as described in Sec. 4-3.7.)

The prevideo signal from the blanking circuit is also applied to a tracking signal generator within IC1101, after the high-frequency components have been removed by Q1113, L1115, C1116, and Q1114. The resultant low-frequency signal is the luminance YL-signal.

The tracking signal generator produces signals that correct for nonlinearity of the pickup tube photoelectric-conversion characteristics in high-luminance sections of the target surface. Such nonlinearity produces a difference output between the Rc/Bc-signals and the luminance signals and results in a condition known as *color shading* or *color blur*.

To prevent color shading or blur, the tracking signal generator produces signals that expand or contract the Rc/Bc-signal according to the luma signal output. The YL-signal is applied to and clipped by three clip circuits. Each clip circuit produces an output tracking signal at pins 5, 7, and 9 of IC1101. These output tracking signals correct the low-luminance, medium-high–luminance, and high-luminance section of the target. The tracking and shading correction circuits are discussed further in Secs. 4-4.5 through 4-4.11.

When a camera subject is under high-luminance light (such as a fluorescent light), colors appear in the highly lighted areas of the target. When light level is low, color shading or blur will stand out. To prevent these conditions, the high/low chroma/clip circuit generates a chroma clip signal that is applied to the automatic white-balance and chroma encoder circuits. The clip circuit, consisting essentially of IC1105 and Q1294 through Q1296, suppresses the chroma when light is above or below normal. The clip level is set by chroma killer level control RT1123 and/or by low-light chroma gain control RT1112.

Note that most prevideo processing circuits do not include both RT1112 and RT1123. We include both on Fig. 4-18 to show possible circuit variations.

IC1105 detects luma signals from the AIC/AGC circuits (input at pin 12) that are above the reference voltage applied to pin 10 and produces a high-luma clip signal at pin 26. This clip signal is added to the tracking signal (through RT1112) and is applied to the noninverting input (emitter) of Q1296. Bias for the tracking signal is applied to the inverting (base) input of Q1296.

Q1296 produces a differential signal that is applied to the automatic

Sec. 4-4   Camera Signal-Processing Circuits (with Pickup Tube)   95

white-balance circuit through Q1295 and 500-ns delay DL1104. When a WHD pulse is applied to clamp Q1294, the differential signal is clamped to about 2.6 V (as set by RT1123 in some processing circuits). DL1104 matches the timing of the clamp output to that of the chroma signal and applies the clip signal to pin 14 of IC1105 (Sec. 4-4.10).

RT1112 controls the level of the tracking signal and sets the level of the chroma suppression under low-luminance conditions. The 1.8 V applied through D1110 to pin 16 of IC1105 is used as a fixed bias.

Note that the output at pin 26 of IC1105 is applied to one section of the white-balance circuits to stop operation when low luminance is detected. The output at DL1104 is applied to another section of the white-balance circuits to inhibit chroma signals under high-luminance conditions.

4-4.3 AIC/AGC Control Circuits

Figure 4-19 shows typical AIC/AGC circuits.

Note that we show two versions of the camera switchboard circuits (manual and gain-up). These circuits represent the two most common iris-control systems.

The circuits provide an AGC (automatic gain-control) function that controls amplifier gain and an AIC (automatic-iris control) function that controls opening and closing of the iris. These two functions combine to control the output level of the video signal.

The AIC detector in IC1101 generates an output signal that corresponds to the average luminance level and applies the signal to the AIC control circuit. In turn, the control circuit generates an output voltage at pin 14 of IC1101, which controls the iris opening.

The AIC circuit detects the Y-signal and generates a d-c level corresponding to the average level of the Y-signal. If luminance information decreases (low light), the AIC control circuit generates an increasing output at pin 14, causing the iris to open. As the iris opens, the amplitude of the video signal increases. When the iris opens all the way, the AGC gain is maximum.

As the light level begins to increase (because the iris is open), the AGC signal maintains a constant video output level by decreasing the AGC voltage. The AGC voltages continues decreasing until the AGC gain is reduced to minimum. At that point, the iris begins closing and controlling the video signal level.

The gain-up circuit (Q1101, Q1102, Q1104) changes the gain of the amplifier, depending on the position of gain-up switch S1601. When S1601 is set to NORMAL, 8 V is applied to the gate of Q1104, turning Q1104 on. This short-circuits feedback resistor R1108 and operates amplifiers Q1101/Q1102 as a buffer.

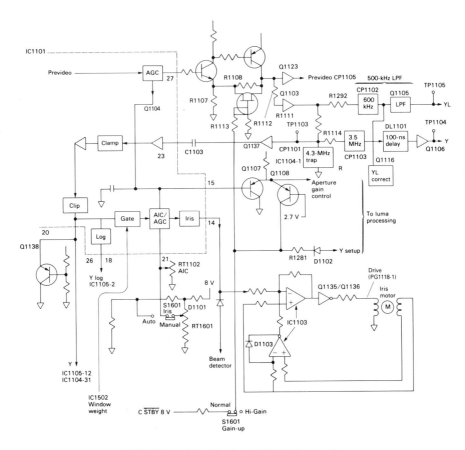

FIGURE 4-19 Typical AIC/AGC circuits

When S1601 is set to HI-GAIN, Q1104 turns off, causing the output of Q1102 to be fed back to Q1101 through R1108. This increases the gain of Q1101/Q1102 by 6 dB.

In the alternate circuit of Fig. 4-19, iris volume control RT1601 provides for manual control of the iris when S1601 is set to MANUAL.

In both versions of the circuit, the AIC control RT1102 sets the balance point of the AIC system.

The AGC voltage from pin 15 of IC1101 is applied to the aperture amplifier in the luma signal-processing circuits (Sec. 4-4.4) through Q1107 and to the AGC circuit within IC1101. AGC voltage increases in proportion to input voltages. So, when AGC voltage becomes higher than the reference voltage (about 2.7 V) at the base of Q1108, Q1108 turns on and applies a high to the gate of Q1104. This turns Q1104 on, short-circuits R1108, and operates Q1101/Q1102 as a buffer.

The prevideo signal is converted to a YL-signal (low-frequency luma signal) by Q1103, CP1102, and Q1105. The prevideo signal is applied through buffer Q1103 to 500-kHz low-pass filter CP1102/Q1105. The low-pass filter removes high-frequency components and passes luma signals (YL-signals) of frequencies less than 500 kHz. To correct the YL level, the R- (red) signal output from pin 1 of IC1104 is added to the YL-signal through Q1116.

The prevideo signal is also converted to a Y-signal by Q1103, CP1101, CP1103, DL1101, and Q1106. The prevideo signal is applied through trap CP1101 where the 4.3-MHz carrier signal is removed. Then CP1103 removes components with frequencies higher than 3.5 MHz.

DL1101 delays the luma signal (Y-signal) by 100 ns to match the timing with that of the chroma signal. The luma signal is also returned to IC1101 (at pin 23) through buffer Q1137. Within IC1101, the luma signal is clamped to 1.8 V and is applied to a high-clip circuit.

Q1138 (at pin 20 of IC1101) generates a clip voltage (about 3.9 V) to clip luma signals in excess of 3.9 V. The clipped luma signals are applied to a log amplifier and to a gate within IC1101. The clipped luma signals are also applied to the prevideo processing circuit (Sec. 4-4.2) and the automatic white-balance circuit (Sec. 4-4.11). The clipped luma signals prevent magenta coloring when the camera is aimed at a high-luminance object and the vertical-edge correction signal is inverted.

The log amplifier within IC1101 amplifies the luma signal. The resultant Y log signal is applied to the vertical-edge correction circuit (Sec. 4-4.6).

The AIC/AGC detector within IC1101 detects the average value of the luma signal level and compares this average with the AIC voltage set by AIC control RT1102, at pin 21 of IC1101. In the alternate circuit, the detector compares the average value with the AIC voltage set by RT1102 and RT1601 (when S1601 is set to MANUAL).

If the luma signal input is lower than AIC voltage, an AGC voltage is produced. If the luma input is higher, a voltage is applied to the iris motor (through IC1103, Q1135, and Q1136), and the iris is closed. The difference between input and AIC voltage determines the resultant output voltage from the detector.

The capacitor within IC1101 holds the AGC voltage (VAGC) and then applies the voltage to the AGC amplifier within IC1101. Since gain of the AGC amplifier is controlled in inverse proportion to the AGC voltage from the detector, the prevideo signal at pin 27 of IC1101 is constant.

As discussed, the AGC voltage is also applied through pin 15 of IC1101 to Q1107/Q1108, which form an aperture gain-control comparator. Q1107/Q1108 match aperture correction level with input level and produce a corresponding signal to the luma signal-processing circuits. This signal (the aperture gain-control voltage) is proportional to input.

Q1107/Q1108 also control the Y setup (or luma setup) signal when gain-up switch S1601 is set to NORMAL. The Y setup signal is applied to the luma signal-processing circuits along with the aperture gain-control voltage.

When the AGC is lower than the reference applied to the base of Q1108 (about 2.7 V), Q1108 turns off, D1102 turns on, and the Y setup voltage decreases. When AGC goes higher than about 2.7 V, Q1108 turns on, D1102 turns off, and the Y setup voltage is determined by the setting of Y setup control R1117 in the luma signal-processing circuits (Sec. 4-4.4).

When the luma signal level is higher than the AIC voltage, an iris voltage proportional to that difference is produced from pin 14 of IC1101. The iris voltage is applied to the iris motor through the motor drive circuits of IC1103, Q1135, and Q1136. If the luma signal increases, the iris voltage increases, the output of IC1103 decreases, and the iris closes. The iris opens if the luma signal decreases.

The iris motor brake coil generates an electromotive force or signal (determined by the position and speed of the motor). This signal is used as negative feedback to IC1103 and causes the iris to open and close smoothly.

As discussed in Sec. 4-3.6, when power is low or when the pickup tube beam is absent or weak, the iris tends to open. To prevent pickup tube burnout under such conditions, beam-detection circuits produce a beam-detection signal to close the iris. When the beam-detection signal through D1101 goes high, the iris is closed completely, overriding any signal from pin 14 of IC1104.

4-4.4 Luminance Signal Processing/Horizontal Aperture Correction Circuits

Figure 4-20 shows typical luminance signal-processing and horizontal aperture-correction circuits.

These circuits process the luma signal (or Y-signal) from Q1106 in the AIC/AGC circuits (Sec. 4-4.3) and apply the processed signal to the chroma encoder (Sec. 4-4.10).

The luma or Y-signal from Q1106 is applied to pin 3 of IC1102. Within IC1102, the signal is clamped to fix the d-c black level, gamma corrected, blanked, dark clipped, and white clipped. The Y-signal from the dark/white clip circuit is branched into two paths. One path is directly to the aperture amplifier within IC1102. The other path is through a 150-ns delay DL1102. The delayed signal is also applied to the aperture amplifier.

The resultant signal output from pin 13 of IC1102 is the low-frequency luminance signal, while the signal from pin 2 of IC1102 is the high-frequency luminance signal (used to emphasize the edges). These two signals are added to produce the aperture-corrected luminance signal.

The feedback clamp within IC1102 fixes the black level (which becomes the signal reference) in the same way as described for the prevideo signal (Sec. 4-4.2). When the BGP pulse is input, the feedback clamp samples the black level and compares the level with the Y setup voltage from the AIC/AGC circuits at pin 8 of IC1102. When the CP1 pulse is input through pin

Sec. 4-4   Camera Signal-Processing Circuits (with Pickup Tube)   99

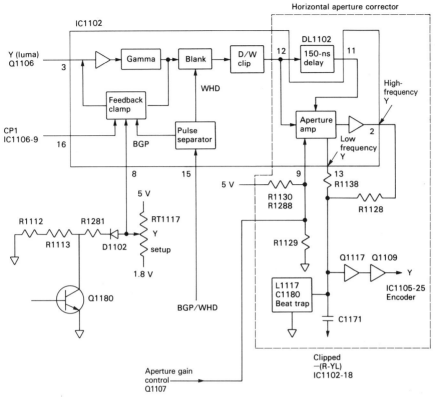

FIGURE 4-20 Typical luminance signal-processing and horizontal aperture correction circuits

16 of IC1102, the differential voltage is fed back, and the input signal is clamped at a fixed d-c level.

Since the voltage on pin 8 of IC1102 is grounded through D1102, R1281, R1113, and R1112, a low-input signal lowers the Y setup voltage. When the input signal level increases and goes higher than the desired level, D1102 turns off. This applies the voltage determined by Y setup control RT1117.

The gamma correction circuit within IC1102 corrects the signal for differences between a pickup tube (Saticon, Newvicon, etc.) and a cathode ray tube.

The blanking circuit within IC1102 uses WHD pulses for blanking in essentially the same way as described for the prevideo signal (Sec. 4-4.2).

The dark/white clip circuit removes (dark clips) synchronous noise in the blanking interval and holds (white clips) the signal level to within rated values. This prevents deterioration in picture quality and minimizes errors in VCR operation.

The horizontal aperture correction portion of the Fig. 4-20 circuit corrects high-frequency signal deterioration caused by pickup tube aperture dispersion. The aperture correction output is created by adding the various signals in the aperture amplifier within IC1102. The output signal is applied to the chroma encoder circuits (Sec. 4-4.10).

The aperture correction circuit receives an input from the AIC/AGC circuit (Sec. 4-4.3) at pin 9 of IC1102, a clipped -(R-YL)-signal from the chroma signal processor circuits (Sec. 4-4.9) and a WHD pulse (to blank the circuit during the blanking interval). The beat trap circuit composed of L1117/C1180 traps beat frequency components (about 720 kHz) between the 4.3-MHz carrier signal and the 3.58-MHz subcarrier signal.

4-4.5 Red/Blue Carrier Separation Circuits

Figure 4-21 shows typical red/blue carrier (Rc/Bc) circuits.

The prevideo signal is applied to the 4.3-MHz bandpass filter CP1105, which passes only the red and blue chroma carrier signals. The resultant chroma signals are then applied through the carrier gain control RT1104 to pin 37 of IC1104 where the signals are amplified.

The amplified signals are applied to an Rc/Bc separation circuit within

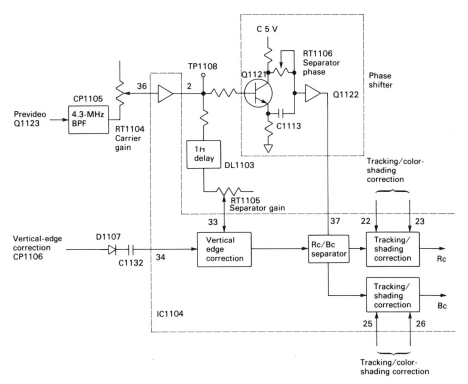

FIGURE 4-21 Typical red/blue carrier (Rc/Bc) circuits

IC1104 through a phase-shifter circuit and to a vertical edge correction circuit within IC1104 through a 1H-delayed circuit. The separated Rc- and Bc-signals are taken from the Rc/Bc circuit and applied to their respective tracking/shading correction circuits (Sec. 4-4.7) within IC1104.

RT1106 and C1113 delay the phase of the Rc/Bc-signal 90° and apply the signals through buffer Q1122 to the Rc/Bc separator within IC1104. DL1103 delays the Rc/Bc-signal by 1H, and RT1105 adjusts the signal gain.

The delayed Rc/Bc-signals and a vertical-edge correction signal (at pin 34 of IC1104) are applied to the vertical-edge correction circuit within IC1104. The internal correction circuit restores the Rc/Bc-signals to their original shape to reject false vertical signals. The resulting signal is applied to the Rc/Bc separator within IC1104.

The Rc/Bc separator adds the 90° phase-delayed signal to the 1H-delayed signal and then subtracts the signals to obtain separate Rc- and Bc-signals. The two separated signals are then applied to the tracking/shading correction circuits.

4-4.6 Vertical-Edge Correction Signal Generator

Figure 4-22 shows typical vertical-edge correction circuits.

The red and blue chroma signals are developed by delaying and adding adjacent horizontal lines. As a result, vertical transitions (white to black or black to white) in the video scene become distorted and usually show cyan or magenta color along the adjacent horizontal lines where the transition occurs.

The vertical-edge circuit improves this situation by detecting the transitions and applying a chroma suppression signal during the period. The correction signal is developed by subtracting the delayed horizontal line from the nondelayed horizontal line to produce a difference signal.

The circuit has one Y input from the AIC/AGC circuits (Sec. 4-4.3) and two outputs. The main output is a chroma suppression signal to the chroma encoder (Sec. 4-4.10). The other output is a correction signal to the red/blue carrier separation circuits (Sec. 4-4.5).

Figure 4-22 shows the signals involved at various points in the circuit. Note that the circuit also uses an SC1 signal from the sync generator (Sec. 4-3.1) and a BGP pulse.

Vertical-edge balance control RT1118 adjusts gain of the Y-signal to match the levels between the undelayed signal and the 1H-delayed signal.

4-4.7 Shading/Tracking Correction Circuits

Figure 4-23 shows typical shading/tracking correction circuits.

Even when dynamic focus is applied to grid 4 of the pickup tube as described in Sec. 4-3.5, the nonuniformity of modulation (due to uneven focus

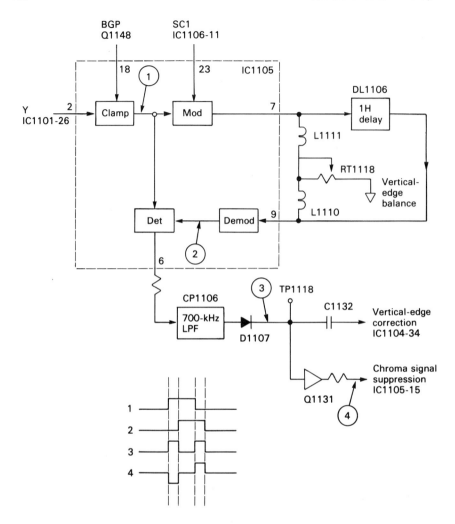

FIGURE 4-22 Typical vertical-edge correction circuits

on the pickup tube target) cannot be completely eliminated. Modulation nonuniformity also occurs due to the nonuniform structure in the photoconductive layer of the pickup tube (manufacturing tolerance).

The shading correction circuit compensates for residual picture shading, using the horizontal and vertical sawtooth and parabolic signals described in Sec. 4-3.4. As shown in Fig. 4-23, these signals are applied to the tracking/shading correction circuits through the color-shading controls RT1509-RT1516 and differential amplifiers in IC1503.

The tracking signals from the tracking signal generator in the prevideo processing circuit (Sec. 4-4.2) are mixed with the shading correction signals

Sec. 4-4   Camera Signal-Processing Circuits (with Pickup Tube)   103

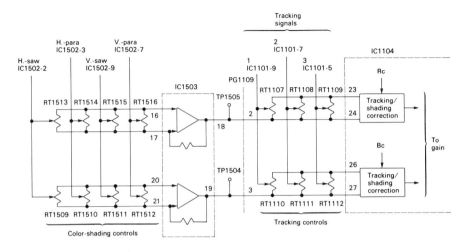

FIGURE 4-23   Typical shading/tracking correction circuits

through the tracking controls RT1107-RT1112. The combined signals are applied to the Rc and Bc shading/tracking correction circuits within IC1104.

4-4.8   Red/Blue Gain-Control Circuits

Figure 4-24 shows typical red/blue gain-control circuits.

These circuits control gain of the Rc and Bc signals to obtain correct white balance and detect the Rc/Bc signals to produce corresponding R and B signals.

The control signals from the automatic white-balance circuit (Sec. 4-4.11) are applied to the gain-control circuits within IC1104 to obtain correct

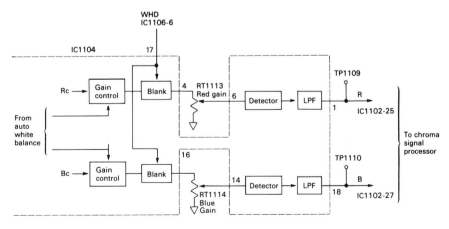

FIGURE 4-24   Typical red/blue gain-control circuits

white balance. The WHD pulse applied through pin 17 of IC1104 blanks the Rc- and Bc-signals. Red gain-control RT1113 and blue gain-control RT1114 set gain of the Rc- and Bc-signals, respectively.

The detector circuits within IC1104 detect the R- and B-signals by removing the 4.3-MHz carrier from the Rc- and Bc-signals. The LPF stages within IC1104 reject any residual 4.3-MHz carrier signal, as well as the high-frequency noise component below about 800 kHz. R and B outputs from the LPF stages are applied to the chroma signal processor circuits (Sec. 4-4.9).

4-4.9 Chroma Signal Processor Circuits

Figure 4-25 shows typical chroma signal processor circuits.

These circuits (contained mostly in IC1102) receive R and B inputs, as well as a low-frequency luminance (YL) input, and generate two color difference signals: -(R-YL) and -(B-YL). The difference signals are applied to the automatic white-balance circuits (Sec. 4-4.11) and the chroma encoder (Sec. 4-4.10). The chroma signal processor circuits also produce a clipped -(R-YL)-signal applied to the horizontal aperture correction circuits (Sec. 4-4.4).

The red or R-signal at pin 25 of IC1102 is clamped to a fixed d-c black

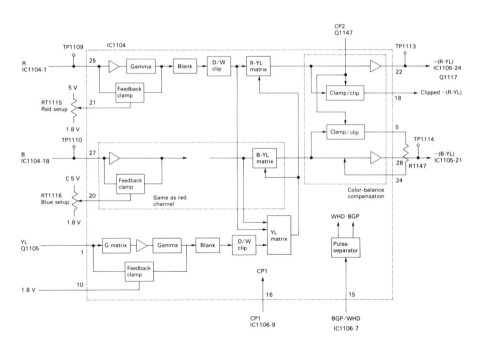

FIGURE 4-25 Typical chroma signal processor circuits

Sec. 4-4   Camera Signal-Processing Circuits (with Pickup Tube)

level by the feedback clamp. The clamp level is set by red setup control RT1115. The clamped signal is amplified and applied to a gamma correction circuit (to correct for differences between a pickup tube and a cathode ray tube).

The gamma correction R-signal is then blanked and applied to a dark/white clip circuit. The output from the dark/white clip circuit is applied to a matrix circuit where the R-, B-, and YL-signals are mixed. The combination of R and YL produces the R-YL-signal.

The blue or B-signal at pin 27 of IC1102 is processed in the same way as the R-signal, except that the clamp level is set by blue setup control RT1116 and the combination of B and YL produces the B-YL-signal.

The YL-signal at pin 1 of IC1102 is first applied to a G-matrix circuit, together with the R- and B-signals. This produces a G- or green signal. The G-signal is then processed in the same way as the R- and B-signals, except that the clamp level is fixed at 1.8 V (and is not adjustable).

The R-YL- and B-YL-signals are applied to a color balance compensation circuit to improve color reproduction. The compensation circuit applies the -(R-YL)-signal to the clamp/clip circuit. When a CP2 pulse is input at pin 6 of IC1102, the clamp/clip circuit clamps the -(R-YL)-signal to about 1.7 V, then clips signals with positive potentials greater than 1.7 V. The clipped signal is output through pin 18 of IC1102 and applied to the horizontal aperture correction circuit to improve color.

The clamp/clip circuit also clamps the -(B-YL)-signal to about 1.3 V and then clips negative-potential signals under 1.3 V. The resultant clipped signal is then passed through R1147 for addition to the -(B-YL)-signal.

4-4.10 Chroma Encoder and Video Processor Circuits

Figure 4-26 shows typical chroma encoder and video processor circuits.

In these circuits, the chroma, luminance, burst, blanking, and sync signals are mixed to produce an NTSC video signal. As described in Chapter 6, the NTSC signal is applied directly to the luminance/chroma signal-processing circuits of the camcorder VCR section. The video signal is also applied to the EVF (Sec. 4-5.3).

The two color difference signals, -(R-YL) and -(B-YL), are applied to clamps within IC1105. When a BGP pulse is input at pin 18 of IC1105, the clamp circuits fix the black level of the color difference signals.

The SC2-signal balance-modulates the -(R-YL)-signal, while the SC1-signal balance-modulates the -(B-YL)-signal. The two modulated color difference signals are then mixed and applied to a chroma killer circuit.

The chroma killer/inhibit circuit receives three inputs to control the level of the color difference signal. The vertical-edge suppression signal suppresses false colors during vertical transitions, as described in Sec. 4-4.6. The high/low-luma/chroma clip signal suppresses chroma signals in high-luma and low-

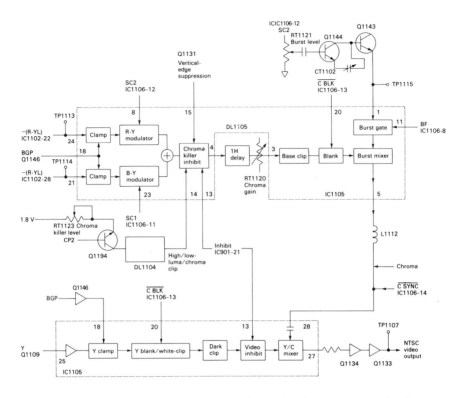

FIGURE 4-26 Typical chroma encoder and video processor circuits

luma sections, as described in Sec. 4-4.2. The inhibit signal from the system-control circuits (Chapter 7) goes high 10 sec after power is applied, and the mode is switch from video to camera, or 1 sec after power-save is released. The inhibit signal suppresses the chroma signal during these intervals.

The chroma killer/inhibit output is applied to a base-clip circuit in IC1105 through external 1H-delayed DL1105 and chroma gain-control RT1120. The base-clip circuit and blanking circuit improve color shading by suppressing low-level chroma signals. The blanking circuit is controlled by a $\overline{\text{C BLK}}$ signal from the sync generator (Sec. 4-3.1).

The burst signal is added to the chroma signal in the burst mixer within IC1105. The burst signal originates as the SC2 signal from the sync generator (Sec. 4-3.1) and is applied to the burst gate in IC1105 through burst-level control RT1121, Q1144/Q1143, and pin 1 of IC1105. The burst gate is on only when the BF pulse from the sync generator IC1106 is high. Control CT1102 sets the phase of the burst signal.

The chroma signal (with burst) is mixed with the Y-signal in the Y/C mixer of IC1105. The sync signal ($\overline{\text{C SYNC}}$ from the sync generator) is added at this point.

When the BGP pulse is input at pin 18 of IC1105, the Y clamp fixes the black level of the Y-signal.

The black/white-clip circuit uses the $\overline{\text{C BLK}}$ pulse from IC1106 to normalize blanking duration and to clip signals that are above normal. The dark-clip circuit clips signals that are below normal to eliminate noise during the blanking period.

When the video inhibit signal from the system-control circuits is high, the video inhibit circuits prevent the Y-signals from passing to the Y/C mixer.

The combined Y-, R-, B-, burst, and sync signals are applied to the VCR luma/chroma circuits of the camcorder and to the EVF (as an NTSC signal) through amplifier Q1134 and buffer Q1133.

4-4.11 Automatic White-Balance Circuit

Figure 4-27 shows typical automatic white-balance control circuits.

These circuits control gain of R- and B-signals so that correct white is produced (white: R = B = G) even when a white object is shot under a variable light source (varying color temperature). Virtually all camcorders have an automatic white-balance function. Many camcorders also have a manual white-balance control. One version of a manual white-balance circuit is shown in Fig. 4-27.

Placing the white-balance switch in AUTO starts the circuit. Placing the white-balance switch in MANUAL stops the circuit, allowing the operator to adjust white balance to any color desired.

When power is first applied, and the 12-V power rises, the reset pulse generator circuit generates a reset pulse (pin 20 of IC1104 goes high). This high resets the automatic white-balance control circuit. Moving the white-balance switch to AUTO applies a high to pin 13 of IC1104 and starts the white-balance start/stop circuit.

A VP pulse from sync generator IC1106 is applied to the start/stop circuit through pin 21 of IC1104. This causes the start/stop circuit to generate a 15-Hz sampling pulse and applies the pulse to the up/down control circuits. The sampling pulse is produced 200 ms after the white-balance switch is set to AUTO.

The clamp circuits clamp the color difference signals at pins 5 and 11 of IC1104 when a CP1 pulse is applied at pin 15. The filter circuits produce an average d-c voltage.

The red and blue comparator circuits compare input voltage (mean value of the color difference signal) with a reference voltage (equal to YL) and produce a signal that is either high or low, depending on the value of the input. The polarity of the comparator outputs is reversed. That is, when -(R-YL) is greater than the reference signal, the red comparator output is high. When -(B-YL) is greater than the reference signal, the blue comparator output is low.

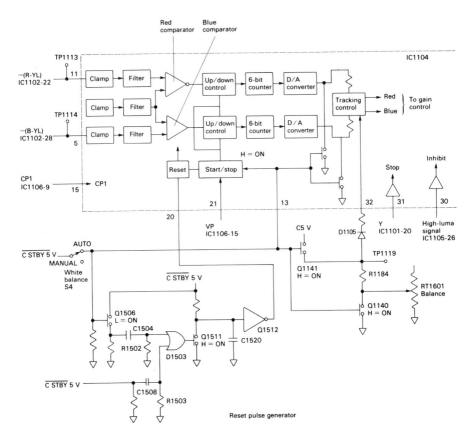

FIGURE 4-27 Typical automatic white-balance control circuits

The 15-Hz sampling pulses applied to the 6-bit counters cause the counters to increment (if the comparator output is high) or decrement (if the comparator output is low).

The 6-bit counters produce corresponding digital data pulses that are converted to analog voltages by the D/A converters. These analog voltages are applied to the tracking-control circuits. In turn, the tracking-control circuit controls gain of the Rc- and Bc-signals, as described in Sec. 4-4.8.

The switch circuit connected to the D/A converter reduces the output impedance of the converters, and thus reduces the width of change in gain-control voltage. The switch circuit is on when the white-balance switch is in AUTO and off when MANUAL is selected.

The low-luminance detector inputs a Y-signal through pin 31 of IC1104 to detect low-luminance conditions and stops operation of the automatic white-balance circuit when luminance drops below a given reference level in IC1104.

Sec. 4-4   Camera Signal-Processing Circuits (with Pickup Tube)   109

The white-balance switching circuit Q1140/Q1141 is turned off when the white-balance switch is set to MANUAL. This connects balance-control RT1601 through R1184 to the tracking-control circuit (in place of the D/A converter outputs) and allows manual adjustment of white balance. When the white-balance switch is set to AUTO, Q1140/Q1141 turn on. This grounds RT1601 and returns the D/A converter outputs to the tracking-control circuit.

The high-luminance inhibit circuit compares the high-luma clip signal at pin 30 of IC1104 (from IC1105, Sec. 4-4.2) with a reference voltage in IC1104. When a high-luminance condition is detected, the circuit produces an inhibit signal that is applied to the filters. This stops operation of the white-balance circuits when luminance goes above a given reference level.

When the white-balance switch is moved from AUTO to MANUAL, Q1506 is turned on, causing the differentiating circuit C1504/R1502 to produce a momentary high pulse. This momentary high is passed through OR-gate D1503 and causes Q1511 to turn on, applying a low to Q1512.

The low from Q1511 is inverted to a high by Q1512 and is applied as a high reset pulse to the reset circuit in IC1104. The reset pulse sets the D/A converter outputs to the center position, so that control of the circuit is entirely determined by balance control RT1601.

4-4.12   Camera Signal-Processing Circuit Troubleshooting

Problems in the camera signal-processing circuits usually show up as picture problems in the EVF and/or monitor. Of course, since the EVF is black and white, you can get a poor color picture on the TV (due to a problem in the color circuits) but a good picture on the EVF. On the other hand, a defect in the black and white circuits (the Y-signal) usually produces problems in both color and black and white performance.

So, before you condemn the signal-processing circuits, check operation on both a TV and the EVF. If the picture appears normal on the TV, the problem is likely in the EVF circuits rather than in the signal-processing circuits. If both the TV and EVF show the same abnormality, suspect the signal-processing circuits.

Finally, before you rip into the signal-processing circuits, make certain that the power distribution circuits (Sec. 4-2) and the deflection circuits (Sec. 4-3) are good.

Pay particular attention to the sync generator circuits described in Sec. 4-3.1. Operation of the signal-processing circuits is dependent on pulses from the sync generator. The same is true of the sawtooth and parabolic signals from the horizontal/vertical sawtooth and parabolic signal generator (Sec. 4-3.4).

If there is no picture, but the EVF deflection appears normal, check that the AIC/AGC circuit of Fig. 4-19 is good. It is possible that the AIC circuit has closed the aperture completely. For example, if the beam-detect signal goes high at D1101, the iris motor closes the iris, as discussed in Sec. 4-4.3.

If you can see the aperture (or hear the iris motor), aim the camcorder at scenes with different light levels and check the response. If this is not practical, monitor the drive voltage to the iris motor and see if there are voltage changes with different light levels. Check adjustment of AIC control RT1102. Also check for proper Y- and YL-signals at TP1104 and TP1105, respectively.

If the Y- and YL-signals are normal (check the service literature for correct waveform and amplitude), and the iris motor is operating with changes in light, it is reasonable to assume that the AIC/AGC circuits are normal. If not, trace signals through the AIC/AGC circuits shown in Fig. 4-19. For example, if there is no drive to the iris motor coil, but the iris drive signal at pin 14 of IC1101 is good, suspect IC1103, Q1135, and Q1136.

Keep in mind that there is a feedback loop in this circuit. For example, if the prevideo signal changes at pin 27 of IC1101, this change is fed back to the AIC/AGC detector in IC1101 and produces a change in the detector output. In turn, the detector change varies the prevideo signal at pin 27 of IC1101.

If the power supply, deflection circuits, and aperture appear normal, but there is no picture, trace the video signal from the pickup tube target to the chroma encoder and video processor. Pay particular attention to the following test points. Check the service literature for correct values at corresponding points on the camcorder you are servicing.

Note that if you have a picture, but there are problems such as color distortion, or color unbalance in a white scene, check for proper alignment of the signal-processing circuits. Chapter 8 gives typical alignment procedures. Always use the service literature for actual procedures.

The video output of the preamp circuit (Fig. 4-17) can be checked at TP1101 and/or TP1102 on the prevideo processing circuits (Fig. 4-18). As a guideline, the video signal at TP1101 is about 0.8 V p-p, while the signal at TP1102 is about 0.3 V p-p, depending on adjustment of RT1103.

The output of the prevideo circuits can be checked at TP1116 (video, typically 0.55 V p-p) and TP1117 (clip signal, also about 0.55 V p-p). Keep in mind that the video signal at TP1116 is dependent on the settings of RT1501–RT1504. The horizontal/vertical sawtooth and parabolic signals (adjusted by RT1501-RT1504) can be monitored at TP1501/TP1502. Typically, the TP1501 signal is about 0.2 V p-p, while the TP1502 signal is about 70 mV p-p. The clip signal at TP1117 depends on the setting of RT1112 or RT1123, whichever configuration is used.

The output of the AIC/AGC circuit (Fig. 4-19) can be checked at the

Sec. 4-4 Camera Signal-Processing Circuits (with Pickup Tube)

iris motor and at TP1103, TP1104, and TP1105. Note that it is in these circuits where the Y, YL, and Rc/Bc are first formed from the prevideo signal.

The Y-signal can be monitored at TP1104, and is about 240 mV p-p. The YL-signal can be monitored at TP1105, and is about 200 mV p-p. The prevideo signal can be measured at TP1103 and is about 0.6 V p-p. Keep in mind that all these signals are dependent on adjustment of RT1102 and on adjustment of RT1117 (Fig. 4-20).

The Y output from the luminance signal-processing circuits (Fig. 4-20) can be monitored at pin 25 of IC1105 (Fig. 4-26) and is about 0.65 V p-p. The Y-signal depends primarily on adjustment of RT1117, at this point.

The combined red/blue carrier signals can be monitored at TP1108 (Fig. 4-21) and are about 0.7 V p-p. Once the carrier signals are separated into Rc and Bc within IC1104, the signals can be measured at pins 4 and 16 of IC1104 or at pins 6 and 14 of IC1104 (Fig. 4-24).

Both Rc- and Bc-signals are dependent on adjustment of RT1105 (Fig. 4-21) as well as adjustment of RT1113/RT1114 (Fig. 4-24). The detected or demodulated R- and B-signals can be monitored at TP1109 and TP1110 and are about 200 mV p-p.

Note that the signals at TP1109 and TP1110 also depend on various correction signals, such as the vertical-edge correction signal applied at pin 34 of IC1104 (Fig. 4-21). The vertical-edge correction signal can be monitored at TP1118 and depends on adjustment of RT1118 (Fig. 4-22).

The signals at TP1109 and TP1110 are also affected by correction signals from the tracking/shading circuit (Fig. 4-23). These signals can be monitored at TP1504 (about 0.6 V p-p) and TP1505 (about 0.9 V p-p).

Finally, the signals at TP1109 and TP1110 are affected by gain-control signals from the automatic white-balance circuit (Fig. 4-27).

To isolate the chroma video signals, first monitor the signal at TP1108 (Fig. 4-21), then at pins 4, 6, 14, and 16 of IC1104, and finally at TP1109/TP1110 (Fig. 4-24).

If the signals are normal at pins 4, 6, 14, and 16 of IC1104, but not at TP1109/TP1110, suspect IC1104, or possibly adjustment of RT1113/RT1114.

If the signals are normal at TP1108, but not at pins 4–16 of IC1104, suspect IC1104, the vertical-edge correction, tracking/shading correction, and automatic white-balance circuits.

If the combined red/blue carrier signals are not normal at TP1108, check adjustment of RT1104 (Fig. 4-21), and then trace back to the prevideo signal at pin 27 of IC1101 (Fig. 4-19).

The output of the chroma signal-processing circuit (Fig. 4-25) can be monitored at TP1113 and TP1114. Both test points should show a signal of about 0.2 or 0.3 V p-p. The signal at TP1113 depends on adjustment of RT1115, while the signal at TP1114 depends on adjustment of RT1116.

The output of the chroma encoder and video processor circuits is (or should be!) a standard NTSC video (1.0 V p-p) signal (complete with chroma, luminance, burst, blanking, and sync). The video output can be monitored at TP1107, in the circuit of Fig. 4-26.

If the signal at TP1107 appears to be normal, the entire camera signal-processing circuits can be considered as good. As a result, some technicians prefer to check here first. Of course, if the output at TP1107 is not normal, you must trace back to the pickup tube, as we have discussed in this section.

4-5 MISCELLANEOUS CAMERA CIRCUITS

In this section, we discuss circuits found in the camera portion of most camcorders. These include the zoom motor control, autofocus, and electronic viewfinder or EVF.

4-5.1 Zoom Motor-Control Circuit

Figure 4-28 shows typical zoom motor-control circuits.

The direction of the zoom motor-drive current is controlled by switches S1602/S1603. Pressing S1602 causes the zoom motor to rotate the lens in the telephoto direction. Pressing S1603 causes the zoom motor to rotate the lens in the wide-angle direction.

The 3.6-V regulator Q1501/ZD1501 on the regulator board supplies power to the zoom switches S1602/S1603.

Troubleshooting for the zoom motor circuits is straightforward. Press each of the switches (S1602/S1603) in turn, and check that the zoom motor moves the lens in and out. If not check wiring between the switches and motor.

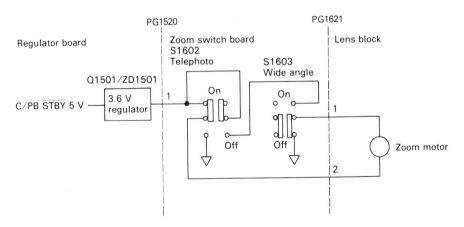

FIGURE 4-28 Typical zoom motor-control circuits

Sec. 4-5 Miscellaneous Camera Circuits 113

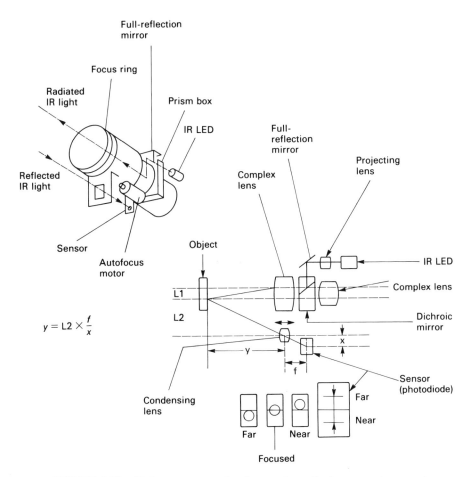

FIGURE 4-29 Main components of an automatic focus system and measurement principles

4-5.2 Automatic Focus Circuit

Figure 4-29 shows the main components of an automatic focus (autofocus) system as well as the measurement principles. Figure 4-30 shows the autofocus circuits in block form. Figure 4-31 shows typical autofocus motor-drive circuits (in a discrete component configuration). In some camcorders, the discrete components shown in Fig. 4-31 are combined within an IC.

As shown in Fig. 4-29, the autofocus control circuit is built into the zoom lens assembly. This circuit uses infrared light (with a center frequency of 870 nm). IR light from an LED is passed through a projecting lens to a full-reflection mirror, where the IR light is reflected back into a dichroic

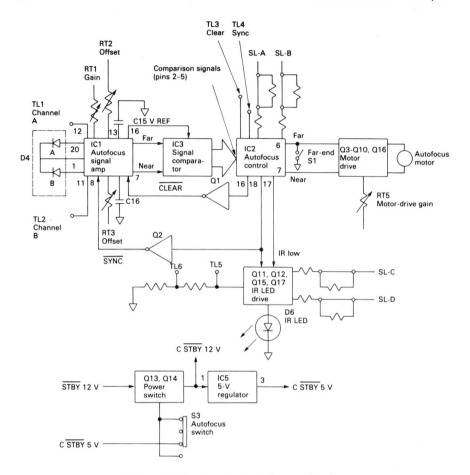

FIGURE 4-30 Typical autofocus circuits

mirror. The dichroic mirror passes visible light (from the scene back to the pickup) but reflects IR light.

The reflected IR light from the dichroic mirror is passed to an object in the scene through a complex lens and is reflected back from the object through a condensing lens to a sensor. The sensor (two IR-sensitive photodiodes) produces a signal that adjusts lens focus so that the reflected IR light is equal on both sensor photodiodes. When this occurs, the lens assembly is properly focused on the object. This technique is generally called the *delta measurement principle* and is essentially the same as that used by a twin-lens reflex rangefinder to obtain the distance between the zoom lens and the object.

When autofocus is selected, power is applied to the circuit of Fig. 4-30. Autofocus microprocessor IC2 produces a 9.1-kHz signal that is applied to

Sec. 4-5   Miscellaneous Camera Circuits   115

Control	Q5	Q6	Q7	Q8	Q9	Q10
Far (pin 6)	On	Off	Off	Off	On	On
Near (pin 7)	Off	On	On	On	Off	Off

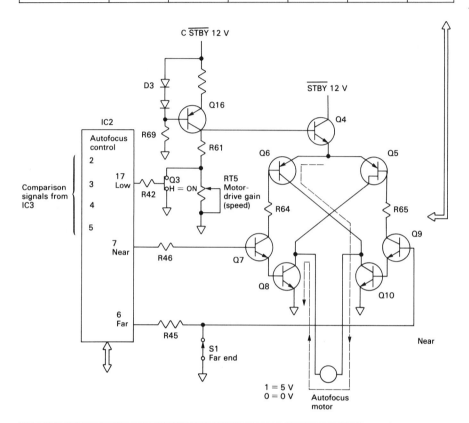

Mode	Input				Output	
	5	4	3	2	6	7
Far	0	1	—	—	1	0
Near	1	0	—	—	0	1
Focused	1	1	0	0	0	0
Far end	1	1	1	1	1	0

FIGURE 4-31 Typical autofocus motor-drive circuits

the IR LED through a drive circuit. The IR signal from the LED is applied to the object, and returned from the object, through the lens assembly.

The reflected IR light is converted to a current by photodiodes A (far) and B (near). The outputs of photodiodes A and B are sync amplified and integrated in IC1.

The integrated far and near outputs from IC1 are compared with a reference voltage from IC1 to generate four comparison signals in IC3. These four comparison signals are applied to IC2 and determine which focus motor-drive direction is required so that the output from photodiode A equals the output of photodiode B. When the outputs are equal, the output circuit turns off, and the focus motor stops.

RT1 sets the gain of IC1, while RT2/RT3 adjust the offset or balance between the A and B channels. RT5 sets gain of the focus motor drive.

Terminals SL-A and SL-B provide a means to short resistors in or out of the IC2 oscillator circuit. This sets the width or period of the sync pulses (typically to 118 ± 15 μs). The sync signal is applied to both the IR LED (through the drive circuit) and to the autofocus signal amplifier IC1.

Terminals SL-C and SL-D provide a means to short resistors in or out of the IR LED drive circuit. This sets the amount of drive current to the LED.

As shown in Fig. 4-31, the focus motor is rotated in the near or far direction by the motor-drive output signal from the autofocus microprocessor IC2. The regulator Q16 generates a constant bias voltage applied to the base of Q4. In turn, Q4 is controlled by motor-speed control RT5. When pin 17 of IC2 goes low, Q3 turns on and shorts RT5 from the circuit. This slows the focus motor speed when proper focus is approached.

When the lens moves to the maximum far direction (infinity), the far end switch S1 is activated mechanically. This shorts R45 to ground, stopping the focus motor.

The chart in Fig. 4-31 shows the status of the motor-drive control transistors Q5-Q10 for both far and near commands. As discussed, the discrete components shown in Fig. 4-31 are combined in a single IC in some camcorders.

Troubleshooting for the autofocus system is discussed in Chapter 9 and is not duplicated here.

4-5.3 Electronic Viewfinder Circuits

Figure 4-32 shows typical EVF circuits in block form.

The EVF displays the video signal from the camera as well as the video signal played back by the VCR section. The EVF is essentially a small-screen black and white monitor. The composite video input signal is received from the camera or the VCR. Most of the EVF circuits are contained within IC1801.

Sec. 4-5   Miscellaneous Camera Circuits   117

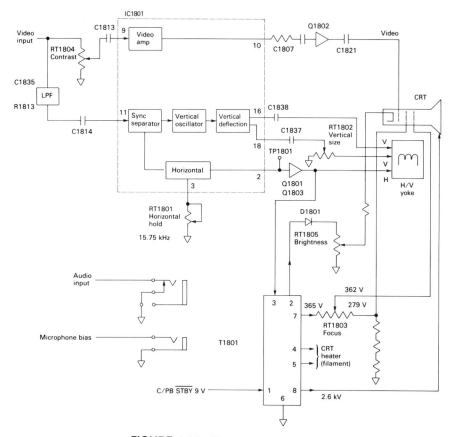

FIGURE 4-32 Typical EVF circuits

The horizontal and vertical sync signals are separated from the composite video signal and applied to the horizontal (H) and vertical (V) deflection circuits. In turn, the H and V deflection circuits provide signals to the horizontal and vertical yokes of the EVF tube. The horizontal deflection circuit also generates the input signal to the high-voltage power supply circuits.

Horizontal-hold control RT1801 sets the frequency of the horizontal oscillator within IC1801 (typically at 15.75 kHz).

Vertical size control RT1802 sets the amplitude of the vertical drive signal from IC1801 to the vertical deflection yoke.

Focus control RT1803 sets the focus voltage from IC1801 to the focus grid of the EVF tube.

Contrast control RT1804 sets amplitude of the video signal to IC1801 and, thus, determines the amount of contrast on the EVF tube.

Brightness control RT1805 sets bias voltage from IC1801 to the cathode and filament of the EVF tube and, thus, determines the amount of brightness on the EVF screen.

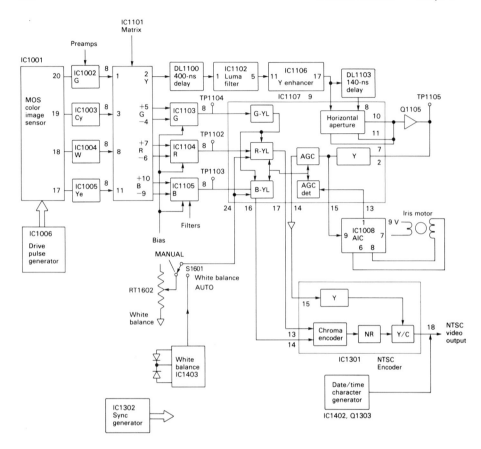

FIGURE 4-33 Signal-processing circuits for camcorders with MOS solid-state pickups or sensors

Troubleshooting for the EVF is discussed in Chapter 9 and is not duplicated here.

4-6 CAMERA SIGNAL-PROCESSING CIRCUITS (WITH MOS PICKUP)

This section describes circuits for camcorders with MOS solid-state pickups (also called *color image sensors*). Only those circuits substantially different from circuits for camcorders with pickup tubes are described.

Figure 4-33 shows the signal-processing circuits for the camera section of camcorders with MOS solid-state pickups or sensors. Compare these circuits with those shown in Fig. 4-16 (for camcorders with pickup tubes).

The MOS *pickup* (*or color image sensor*) IC1001 develops four color signals—cyan (Cy), green (G), yellow (Ye), and white (W)—when driven

Sec. 4-6   Camera Signal-Processing Circuits (with MOS pickup)   119

by signals from the *drive pulse generator* IC1006. The four signals are applied to *preamplifier* ICs IC1002-IC1005.

The preamplifiers amplify the color signals to form prevideo signals. In turn, the prevideo signals are applied to *matrixing circuit* IC1101.

IC1101 converts the four color signals (G, Cy, Ye, and W) into the luma signal (Y) and chroma signals (R, B, and G). The luma and chroma signals are applied to corresponding *filters* IC1102-IC1105.

Delay line DL1101 delays the luma signal by 400 ns to match the phase with the chroma signals. DL1101 also eliminates the 5.43-MHz horizontal shift-register clock pulse. *Luma filter* IC1102 buffers and blanks the luma signal. The output from IC1102 is applied to *Y enhancer* IC1106, where the vertical edges of the picture are emphasized.

The *luma/chroma processing circuits* IC1107 and DL1103 perform two basic functions. First, the *horizontal aperture correction circuit* corrects horizontal aperture. The corrected luma or Y-signal is then processed for application to the NTSC encoder IC1301. Second, the *chroma-processing circuits* process the chroma signals (R, B, G, YL) and generate two color difference signals R-YL and B-YL. The color difference signals are applied to the chroma encoder in IC1301.

The NTSC encoder IC1301 combines the chroma and luma signals in a Y/C mixer to produce a standard NTSC signal. This signal is combined with that of the date/time character generator.

The *date/time character generator* IC1402/Q1303 generates date and time character signals (to be displayed on the EVF and recorded on tape). The date/time signals are mixed with the NTSC video from IC1301.

Sync generator IC1302 generates sync pulses and scan pulses for processing the signal, and for synchronization with the MOS drive pulse generator IC1006.

The *AIC circuit* IC1108 controls the iris opening (as determined by the level of the Y-signal).

The *automatic white-balance control circuit* IC1403 detects the color temperature (by sensing red and blue light) and adjusts the red/blue gains accordingly to maintain the correct white balance.

4-6.1 MOS Color Image Sensor (pickup) Circuits

Figure 4-34 shows the solid-state structure and basic circuit configuration of a typical MOS color image sensor or pickup. Figure 4-35 shows a typical circuit configuration.

The sensor shown in Figs. 4-34 and 4-35 has an image size of 8.8 mm (H) × 6.5 mm (V) to match a $\frac{2}{3}$-in. optical system. Chip size is 9.9 mm (H) and 8.0 mm (V). The number of picture elements is 576 (H) × 486 (V), with a total of approximately 280,000.

With this particular sensor (the HE98241), sensitivity in the infrared

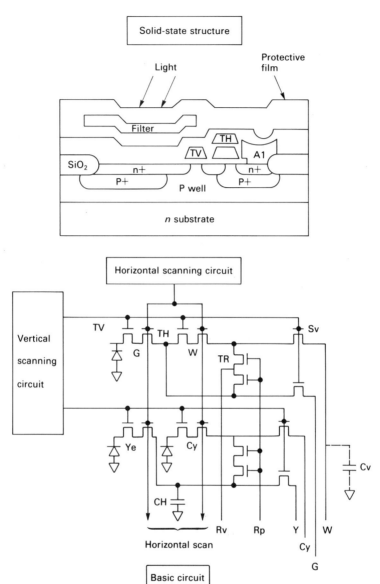

FIGURE 4-34 Structure and basic circuit of a typical MOS color image sensor or pickup

region is low. Well-balanced spectral sensitivity (good color representation across the color spectrum) is obtained by use of picture elements (photodiodes) with a three-layer NPN structure. This solid-state structure also suppresses "blooming" of the picture (a condition common in camcorders with pickup tubes).

Sec. 4-6 Camera Signal-Processing Circuits (with MOS pickup)

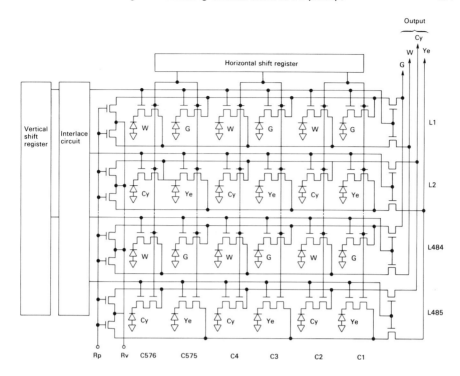

FIGURE 4-35 Typical circuit configuration for MOS sensor

The color resolution filter (that produces the four basic colors) is made by arranging complementary white, yellow, cyan, and green color filters in a mosaic.

The picture elements are driven by signals from the drive pulse generator (Sec. 4-6.3).

The four color signals—W, Ye, Cy, and G—are read out on four output lines. This provides for interlaced scanning with high resolution and without residual image.

The relationship between scanning and picture elements to produce the color signals is described next.

As shown in Fig. 4-34, when light falls on a picture element, an electron/hole pair is generated inside the $n+$ layer and $p+$ layer, which form the photodiode. As the electron in the $p+$ layer flows out to the $n+$ layer, a hole remains. As the hole in the $n+$ layer flows out to the $p+$ layer, an electron remains. The resultant photoelectrons are stored in the $n+$ layer as a photoelectronic conversion signal.

When the vertical scanning pulses open the gates of the TVs, and the horizontal scanning pulses open the gates of the corresponding THs (in sequence from left to right), the photoelectrons of the photodiodes of two horizontal lines are output in sequence. When the gate opened by the vertical

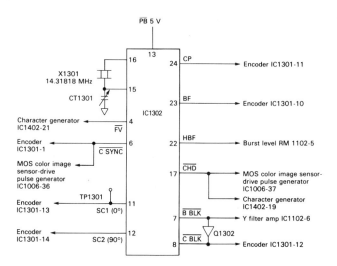

FIGURE 4-36 Sync generator circuits for camcorder with MOS sensor

scanning circuit is changed in sequence, and repeated, the photoelectrons of all photodiodes are output in sequence, and all picture elements are scanned.

The four signals (Ye, Cy, G, and W) are output simultaneously by a 5.43-MHz sampling signal from the drive pulse generator (Sec. 4-6.3). The signals are mixed in the matrix circuit (Sec. 4-6.5) after preamplification (Sec. 4-6.4) to produce the luminance (Y-) signal as well as the chroma signal (R-, B-, and G-).

4-6.2 Sync Generator Circuits

Figure 4-36 shows the sync generator circuits used in a typical camcorder with MOS sensor. Compare this circuit with the sync generator described in Sec. 4-3.1 and shown in Fig. 4-7.

4-6.3 Sensor-Drive Pulse Generator Circuits

Figure 4-37 shows the drive pulse generator circuits for a typical MOS color image sensor.

IC1006 is a CMOS IC that produces the pulses to drive the MOS color image sensor IC1001, as well as the pulses to process the signal. The circuit configuration of IC1006 is roughly classified into three sections: a 5.43-MHz high-speed section, a horizontal frequency section, and a vertical frequency section. IC1006 synchronizes the horizontal counter with the $\overline{\text{CHD}}$ from sync generator IC1301 and operates the vertical counter by detecting the field using the $\overline{\text{C SYNC}}$ and $\overline{\text{CHD}}$ pulses from IC1301.

The 5.43-MHz high-speed section generates the horizontal shift-register

Sec. 4-6   Camera Signal-Processing Circuits (with MOS pickup)   123

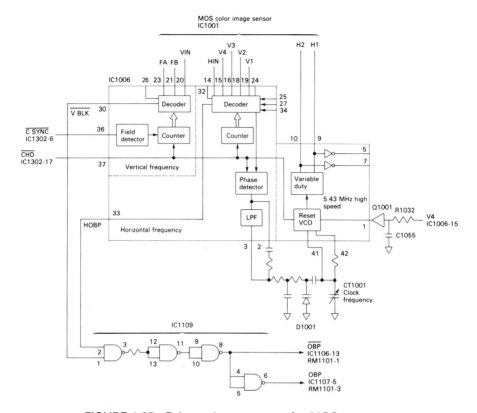

FIGURE 4-37   Drive pulse generator for MOS sensor

clock pulses H1 and H2. The 5.43-MHz frequency of H1 and H2 is determined by the number of picture elements in IC1001.

The phase detector detects the phase differences between the $\overline{\text{CHD}}$ and reference pulses from the horizontal decoder and generates an oscillator frequency control voltage through a low-pass filter. This voltage exits IC1006 at pin 3 and reenters IC1006 at pin 2.

The control voltage is applied to the voltage-controlled oscillator (VCO), which is reset every 1H by a reset pulse to cancel phase fluctuation at the fall (start of horizontal blanking) of the $\overline{\text{CHD}}$ pulse. The reset pulse is generated by the phase difference between the $\overline{\text{CHD}}$ pulse and the vertical buffer pulse V4.

Because phase fluctuations are present in the horizontal frequency, the horizontal start pulse changes every 1H, causing horizontal jitter. The VCO generates a 5.43-MHz sinewave signal after being reset every 1H. The 5.43-MHz signal is applied to the variable duty circuit and the H1 and H2 pulses are generated.

The VCO is set to the correct clock frequency by CT1001.

The horizontal frequency section generates the following horizontal frequency (15.734 kHz) pulses, using the horizontal shift-register clock pulse H2 as the reference:

(HIN)	Horizontal shift-register start pulse	Pin 14
(V4)	Vertical buffer pulse	Pin 15
(V3)	Vertical buffer pulse	Pin 16
(V2)	Vertical shift-register clock pulse	Pin 18
(V1)	Vertical shift-register clock pulse	Pin 19
(HOBP)	Horizontal optical black pulse	Pin 33

The vertical frequency section generates the following vertical frequency (60-Hz) pulses, after receiving the $\overline{\text{CHD}}$ and C SYNC pulses:

(VIN)	Vertical shift-register start pulse	Pin 20
(FB)	Field discrimination pulse	Pin 21
(FA)	Field discrimination pulse	Pin 23
(V BLK)	Vertical blanking pulse	Pin 30

The optical black pulse generator generates OBP and $\overline{\text{OBP}}$ pulses after inputting HOBP and $\overline{\text{V BLK}}$ from IC1006.

4-6.4 Preamplifier circuits

Figure 4-38 shows the preamplifier circuits used in a typical camcorder with MOS sensor.

The four signals Cy, G, Ye, and W from the MOS color image sensor IC1001 are applied to the preamplifier circuits in IC1002-IC1005. The output signal of IC1001 is a constant current supply and is very weak (about 500 nA) when compared to a pickup tube. As a result, the signal-to-noise (S/N) ratio of camcorder with MOS sensors is determined largely by the preamplifiers.

Preamplifiers IC1002-IC1005 input the corresponding signal during the 5.43-MHz (174-ns) clock period. Negative feedback within the preamplifier ICs keeps the input impedance low. The voltage feedback to the gate of the FET amplifier within the preamplifiers becomes the video bias for the MOS color image sensor. The bias applied to the Cy, Ye, G, and W signals is about 2.5 V. Output signals from the preamplifiers are applied to the matrixing circuits in IC1101.

4-6.5 Matrixing and Filter Circuits

Figure 4-39 shows the matrixing and filter circuits used in a typical camcorder with MOS sensor.

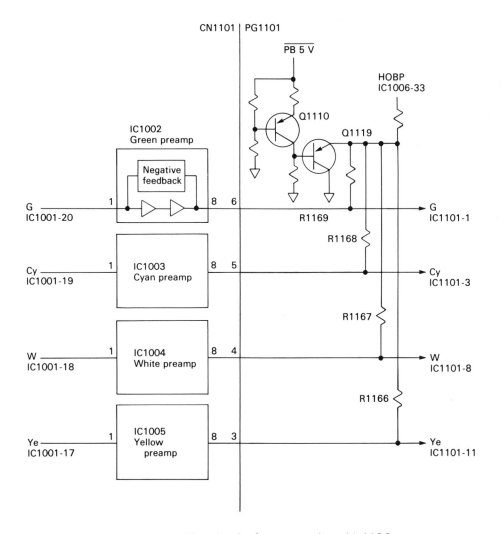

FIGURE 4-38 Preamplifier circuits for camcorder with MOS sensor

The Cy, G, Ye, and W signals from the preamplifiers IC1002-IC1005 are applied to the matrixing circuits in IC1101 and are matrixed to generate the luma (Y-) and chroma signals (R-, B-, and G-). The signals are amplified (differentially) by the filters IC1102-IC1105 as follows:

$$Y = Cy + G + Ye + W$$
$$= (B + G) + G + (R + G) + (R + B + G)$$
$$= 2R + 2B + 4G$$
$$G = (Cy + Ye + G) - W$$
$$= (B + G + R + G + G) - (R + B + G)$$
$$= 2G$$

125

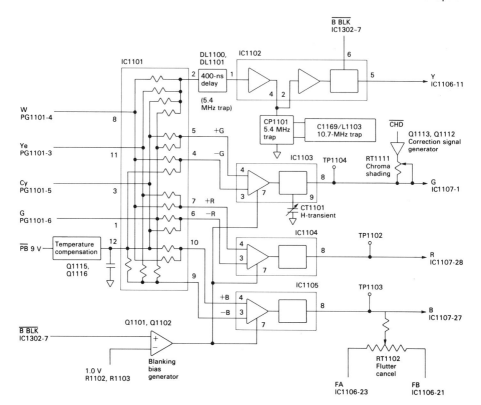

FIGURE 4-39 Matrix and filter circuits for camcorder with MOS sensor

$$R = (Ye + W) - (Cy + G)$$
$$= (R + G + R + B + G) - (B + G + G)$$
$$= 2R$$
$$B = (Cy + W) - (Ye + G)$$
$$= (B + G + R + B + G) - (R + G + G)$$
$$= 2B$$

The Y-signal output from pin 2 of IC1101 is subjected to a 400-ns delay by DL1100/DL1101. This matches the phase of the Y-signal with that of the chroma signals. The 5.43-MHz horizontal shift-register clock pulse is also eliminated at this point.

The delayed Y-signal is amplified by the amplifier in IC1102, and any remaining 5.43-MHz pulse is eliminated by trap CP1101. The 10.7-MHz trap circuit C1169/L1103 is connected in parallel with CP1101 and eliminates the second harmonic component of the 5.43-MHz pulse.

The Y-signal reenters IC1102 at pin 2 and is amplified. After amplification, the Y-signal is blanked by the $\overline{\text{B BLK}}$ pulse from IC1302 and is applied to the Y enhancer IC1106.

The G-signal is generated when the +G- and −G-signals are amplified (differentially) in IC1103. The G-signal is blanked by applying a bias voltage form Q1101/Q1102 at pin 7 of IC1103.

G-signal components above 700 kHz are eliminated by the low-pass filter (LPF) within IC1103. G-signal components below 700 kHz are applied to chroma processor IC1107.

The H-transient trimmer CT1101 at pin 9 of IC1103 controls the frequency characteristics of the LPF to match the phase with other chroma signals. This is done to minimize *horizontal smear*.

The $\overline{\text{CHD}}$ pulse is integrated by Q1113/Q1112 and is added to the G-signal through chroma shading control RT1111. This makes it possible to adjust for improved *horizontal color shading*.

The R-signal is generated when the +R- and −R-signals are amplified (differentially) in IC1104. The R-signal is blanked by applying a bias voltage from Q1101/Q1102 at pin 7 of IC1104. R-signal components below 700 kHz are applied to chroma processor IC1107.

The B-signal is generated when the +B- and −B-signals are amplified (differentially) in IC1105. The B-signal is blanked by applying a bias voltage from Q1101/Q1102 at pin 7 of IC1105. B-signal components below 700 kHz are applied to the chroma processor IC1107.

FA and FB pulses are added to the B-signal through flutter cancel control RT1102. This makes it possible to adjust for minimum *flutter noise*.

The bias voltage applied to differential amplifiers within IC1103-IC1105 is generated by Q1101/Q1102. The $\overline{\text{B BLK}}$ pulse applied to the base of Q1101 is buffered and then limited to be higher than the voltage of about 1.6 V generated by Q1102.

4-6.6 Y Enhancer Circuits

Figure 4-40 shows the Y enhancer circuits used in a typical camcorder with MOS sensor.

The Y-signal from filter IC1102 is clamped within IC1106 to a specified voltage when the $\overline{\text{OBP}}$ pulse is applied. The clamped Y-signal is then processed by two filters and an adder within IC1106.

A low-pass filter is formed by amp 1 and the CCD 1H delay in IC1106, together with external LPF DL1102. The 1H delay is driven by a 10.7-MHz clock pulse and delays the Y-signal by about 1H. The delayed Y-signal is further delayed by DL1102, so that the total delay is exactly 1H.

The 1H-delayed Y-signal reenters IC1106 at pin 20 and is applied to amp 1 (an op-amp). The output from amp 1 is applied to amp 2 (another op-amp) through a 1-k gain-control circuit (to match the gains of amp 1 and amp 2).

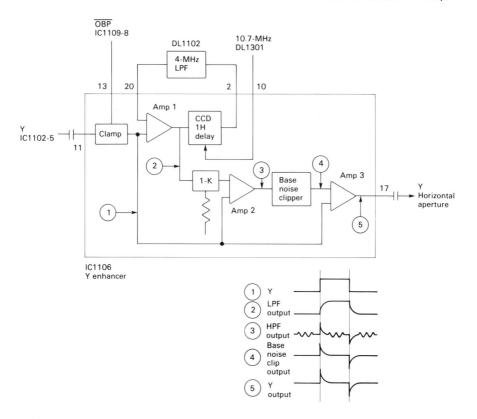

FIGURE 4-40 Y enhancer circuits for camcorder with MOS sensor

A high-pass filter is formed by amp-2 and the base noise clipper in IC1106. In amp 2, the low-pass filtered Y-signal (2) is subtracted from the original Y-signal (1) to produce an edge-emphasizing signal (3) at the output of amp 2. The edge-emphasizing signal (3) is applied to the base noise clipper to eliminate any remaining high-frequency noise, leaving only the clipped edge-emphasizing signal (4). The clipped signal (4) is then applied to the adder.

An adder circuit is formed by amp 3 (an op-amp). The original Y-signal (1) and the clipped signal (4) are added in amp 3 to produce a vertical edge–emphasized Y-signal (5) that is applied to the horizontal aperture correction circuit in IC1107.

4-6.7 Horizontal Aperture Correction/Luma Signal-Processing Circuits

Figure 4-41 shows the horizontal aperture correction and luma signal-processing circuits used in a typical camcorder with MOS sensor.

Sec. 4-6 Camera Signal-Processing Circuits (with MOS pickup) 129

The horizontal aperture correction circuit is composed primarily of the 140-ns delay line DL1103 and the aperture circuit within IC1107. The Y-signal from the Y enhancer (with the vertical edge emphasized) is branched into two parts. One part is applied through pin 9 of IC1107 to the aperture circuit, composed of an adder and subtractor. The other part is applied to the DL1103. The output of DL1103 is also applied to the adder and subtractor of the aperture circuit. Note that since the adder and subtractor inputs are not terminated, the input (3) at pin 9 of IC1107 becomes the nondelayed (1) and twice-delayed (2) signals which are added, as shown by the waveforms on Fig. 4-41.

In the adder, signals 3 and 4 are added to product signal 5 output at pin 11 of IC1107. In the subtractor, signal 4 is subtracted from signal 3 to produce signal 6 output at pin 7 of IC1107.

A base noise clip function in the aperture circuit eliminates any remaining high-frequency noise in signal 6. Signals 5 and 6 are combined and applied through buffer Q1105 to produce the horizontal aperture corrected Y-signal 7. A CHD pulse from Q1112 is applied to Y-signal 7 through luma shading control RT1110. This is done to improve horizontal color shading.

The luma signal-processing circuit components are found mostly in IC1107. The corrected Y-signal (7) is applied to the luma circuits at pin 2 of IC1107. The d-c black level of the corrected Y-signal is adjusted by Y setup control RM1101-4.

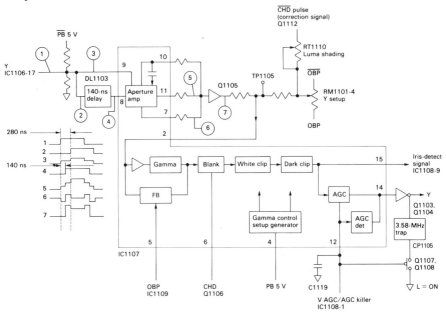

FIGURE 4-41 Horizontal aperture correction and luma signal-processing circuits for camcorder with MOS sensor

The Y-signal is clamped to a fixed d-c level by the FB (feedback) clamp within IC1107. This prevents variation in signals applied to the gamma correction circuit. The black level of the gamma-corrected Y-signal is sampled and held by a capacitor when an OBP pulse is applied. This sample voltage is compared with a setup voltage generated in IC1107, and the difference is applied to the gamma correction circuit.

The gamma correction circuit within IC1107 corrects the signal for differences between the MOS sensor and a cathode ray tube.

The blanking circuit within IC1107 uses CHD pulses to eliminate noise in the blanking period.

The Y-signal is dark clipped and white clipped (in a linear fashion) by the dark-clip and white-clip circuits, respectively. This sets the level of the Y-signal to a rated value and eliminates any noise remaining in the blanking period. The clipped output is branched into two parts. One part is applied through pin 15 of IC1107 to the automatic iris-control (AIC) circuit (Sec. 4-6.8), while the other part is applied through the AGC amplifier to the AGC detector.

The AGC circuit detects the level of the Y-signal by comparison to a reference voltage and generates an inverse AGC voltage that corresponds to the difference. The AGC voltage is made constant by C1119 and is applied to the AGC amplifier in IC1107. As usual, the AGC voltage increases when AGC output decreases, and vice versa. As a result, the output from the AGC amplifier is constant.

The output from the AGC amplifier at pin 14 of IC1107 is applied through buffer Q1103/Q1104 to the video processing circuit IC1301 (Sec. 4-6.11). An AGC killer voltage from the AIC circuit IC1108 (Sec. 4-6.8) is applied at pin 12 of IC1107. This killer voltage suspends AGC operation and sets the gain of the AGC amplifier to minimum until the iris is fully open.

The 3.58-MHz trap circuit CP1105 is controlled by the level of the AGC voltage present at pin 12 of IC1107. The circuit reduces high-frequency noise under low-light conditions. When illumination decreases, the AGC voltage also decreases. When the AGC voltage decreases to about 2.8 V, both Q1107 and Q1108 turn on, and CP1105 is grounded.

4-6.8 Automatic Iris-Control (AIC) Circuits

Figure 4-42 shows the AIC circuits used in a typical camcorder with MOS sensor.

These circuits provide both an AGC function that controls amplifier gain and an AIC function that controls opening and closing of the iris. The two functions combine to control the output level of the video signal.

The Y-signal from IC1107 (Sec. 4-6.7) is applied to a filter composed of C1110/R1127 as the iris-detect signal. The average level of the Y-signal is applied to a differential amplifier within IC1108. The average level of the

Sec. 4-6   Camera Signal-Processing Circuits (with MOS pickup)   131

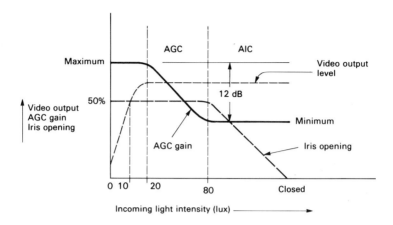

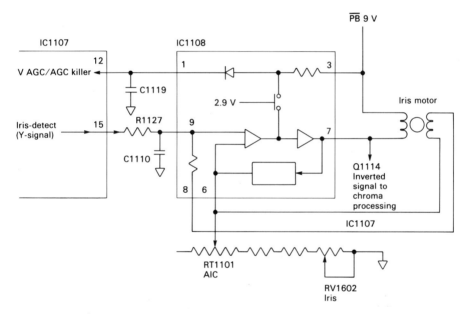

FIGURE 4-42 AIC circuits for camcorder with MOS sensor

Y-signal is amplified, based on the reference voltage at the inverting input. The reference voltage is a combination of feedback from the iris motor and the voltage set by AIC control RT1101 and IRIS control RV1602. Typically, RT1101 is an internal adjustment, whereas RV1602 is an operating control.

When the average lighting of the scene viewed by the camera decreases, the Y-signal output also decreases, and vice versa. The iris motor is operated by the voltage difference between the 9-V source and the output of the motor driver at pin 7 of IC1108. When the average lighting increases, the motor driver output increases, and the iris closes. The iris opens when average lighting decreases. The feedback signal developed by the brake coil of the iris motor is applied to both inputs of the differential amplifier to control iris motor operation.

The AGC killer voltage from pin 1 of IC1108 is applied to the AGC amplifiers in the Y-signal circuits described in Sec. 4-6.7. The output of the differential amplifier in IC1108 is applied through a switch circuit to pin 1. When the average lighting increases, the output of the differential amplifier increases. If this output goes higher than about 2.9 V, the switch circuit turns off, causing the diode to turn on, and applies a high to IC1107 through pin 1. This sets AGC gain (in IC1107) to minimum. When the output of the differential amplifier is lower than 2.9 V, the switch circuit goes on, the diode turns off, and the AGC amplifiers in IC1107 assume control of the Y-signal (and, thus, control of the video signal).

The crossover points for AIC/AGC control are shown in Fig. 4-42.

4-6.9 Chroma Signal-Processing Circuits

Figure 4-43 shows the chroma signal-processing circuits used in a typical camcorder with MOS sensor.

The G-signal from the G filter/amplifier IC1103 (Sec. 4-6.5), the R-

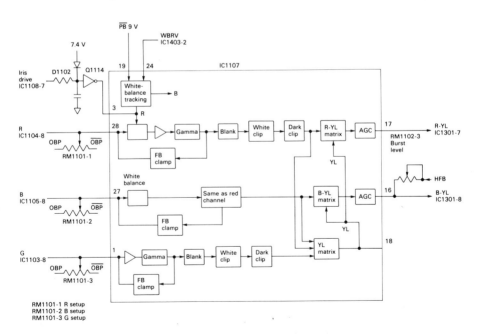

FIGURE 4-43 Chroma signal-processing circuits for camcorder with MOS sensor

Sec. 4-6 Camera Signal-Processing Circuits (with MOS pickup)   133

signal from the R filter/amplifier IC1104, and the B-signal from the B-filter/amplifier IC1105 are applied to the chroma signal-processing circuits in IC1107 which generates two color different signals R-YL and B-YL.

In the G-signal-processing circuit, the d-c level of the G-signal is set to the rated level by adjusting the G setup control RM1101-3 (in the same way as the Y-signal, Sec. 4-6.7). The black level of the G-signal is clamped to a fixed d-c level by the FB clamp within IC1107. This prevents variation in signals applied to the gamma correction circuit (which corrects the signal for differences between the MOS sensor and a cathode ray tube). The blanking circuit uses CHD pulses to eliminate noise in the blanking period. The G-signal is dark/white clipped to set the level and eliminate any remaining noise.

To generate the YL-signal, the G-, R-, and B-signals are matrixed in IC1107. The YL-signal is applied to color difference (R-YL and B-YL) signal-matrixing circuits.

The R-signal is processed in the same way as the G-signal, except for the white-balance, color difference signal-matrixing circuit and the AGC amplifier.

The white-balance circuit in the R-signal path is an AGC amplifier, the gain of which is controlled by the red control voltage from the white-balance tracking circuit. The gain-control characteristics are mutually compensated (seesaw) in the R and B channels to maintain the correct white balance. In the R channel, the output is proportional to the level of the white-balance control voltage (WBRV) from the automatic white-balance circuits of IC1403 (Sec. 4-6.10).

When the camera is pointed at a low-brightness object, and the iris drive voltage applied from the AIC circuits of IC1108 (Sec. 4-6.8) drops to open the iris, Q1114 turns on. The voltage at the collector of Q1114 boosts the red control voltage applied to pin 3 of IC1107. This raises the gain of the R channel white-balance control circuit.

The R- and YL-signals are matrixed to generate the color difference signal R-YL.

The gain of the AGC circuit is controlled in the same way as the Y-signal (Sec. 4-6.7). The output is inversely proportional to the level of the AGC voltage. The R-YL output signal leaves IC1107 at pin 17 and is applied to the encoder IC1301 as described in Sec. 4-7.11.

The B-signal is processed in the same way as the R-signal, and the resultant B-YL-signal is applied to the encoder IC1301. However, the burst level of the B-YL-signal is determined by an HBF pulse applied through burst level control RM1102-3.

4-6.10 Automatic White-Balance Control Circuits

Figure 4-44 shows the automatic white-balance control circuits used in a typical camcorder with MOS sensor.

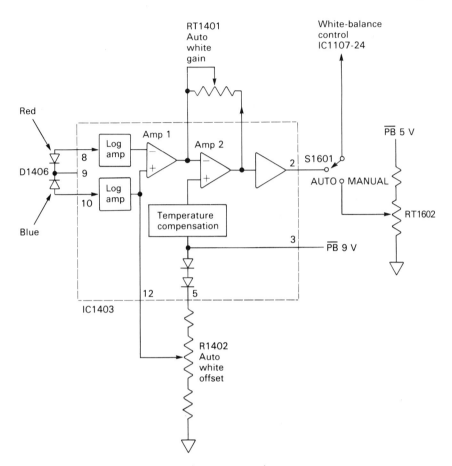

FIGURE 4-44 Automatic white-balance control circuits for camcorder with MOS sensor

This circuit is responsible for controlling gain of the red and blue chroma signals to maintain correct white balance (color temperature) under various lighting conditions. This ensures that white areas of the picture will not have a red or blue cast or tint.

When S1601 is set to MANUAL, the white balance control RT1602 is connected to the white balance tracking circuit in IC1107 as discussed in Sec. 4-6.9. RT1602 provides manual control of white balance by adjustment of the WBRV voltage.

When S1601 is set to AUTO, the output of the circuit in IC1403 is connected to the white-balance tracking circuit in IC1107, and correct white balance is maintained automatically.

The color balance detector detects the red and blue spectra by means

Sec. 4-6   Camera Signal-Processing Circuits (with MOS pickup)

of corresponding photodiodes in D1406. The red and blue photodiodes generate currents corresponding to the levels of the colors. The currents are applied to respective logarithmic amplifiers in IC1403. Outputs from the log amplifiers are applied to amp 1 (an op-amp) which generates a d-c voltage that corresponds to the *ratio of the red and blue spectra* (or color temperature). The auto white offset control RT1402 at pin 12 of IC1403 controls the offset voltage of amp 1 and, thus, makes it possible to adjust for a given white balance under given conditions of color temperature.

The output of amp 1 is applied to the inverting input of amp 2. The noninverting input of amp 2 receives a reference signal from a temperature-compensation circuit in IC1403. The output of amp 2 is buffered and applied to the white-balance tracking circuit in IC1107. The auto white gain control RT1401 controls gain of amp 2 and, thus, makes it possible to adjust for the correct amount of white-balance control.

The temperature-compensation circuit has a negative temperature coefficient, so when the temperature changes, the white-balance output voltage (WBRV) remains constant. (When temperature drops, the compensation circuit generates an increased voltage to amp 2, and vice versa.)

4-6.11 Encoding Circuits

Figure 4-45 shows the encoding circuit used in a typical camcorder with MOS sensor.

Most of the components for this circuit are contained within encoder IC1301. The Y-signal from the luma signal-processing circuits (Sec. 4-6.7) is applied at pin 6 of IC1301. The -(R-Y) and -(B-Y) color difference signals from the chroma signal-processing circuits (Sec. 4-6.1) are applied at pins 7 and 8 of IC1301, respectively. The color difference signals -(R-Y) and -(B-Y) are mixed (or encoded) to generate a chroma signal. In turn, the chroma signal is mixed (encoded) with the Y-signal to produce the NTSC video signal at pin 3 of IC1301.

The Y-signal is clamped to fix the d-c level when a CP pulse is applied. The clamped Y-signal is blanked during the blanking period to eliminate noise and is white-clipped to reduce the color output in very bright areas of the picture. The blanked/clipped Y-signal is dark-clipped to reduce the color output amplitude in very dark areas of the picture and to minimize any noise remaining in the blanking period. The processed Y-signal is then combined with the chroma signal in the Y/C mixer and leaves IC1301 at pin 3 as an NTSC signal.

The -(R-YL)-signal is added to the $\overline{\text{C SYNC}}$ pulse, and the C SYNC pulse (inverted by Q1301) in carrier leak 1 controls RT1301 to cancel any carrier leak. The -(R-Y)-signal (with carrier leak removed) is applied through a buffer at pin 7 of IC1301 and is clamped to fix the d-c level when a CP

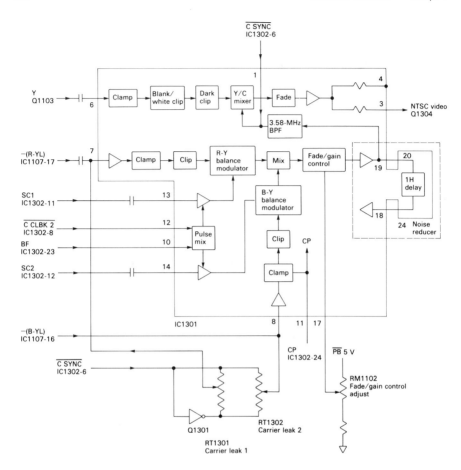

FIGURE 4-45 Encoding circuit used in camcorder with MOS sensor

pulse is applied. A clip circuit then clips the signal to a rated level. The clipped -(R-Y)-signal is then balance modulated by the SC1 (0° phase-shifted 3.58-MHz subcarrier) in the R-Y balanced modulator and applied to a mixer.

The -(B-YL)-signal is applied to the mixer in the same way as is the -(R-YL)-signal, except that carrier leak 2 control RT1302 is used to cancel carrier leak and the -(B-YL)-signal is balance modulated by the SC2 (90° phase-shifted 3.58-MHz subcarrier).

The $\overline{\text{C BLK}}$ pulse from pin 12 of IC1301 and the BF pulse from pin 10 are mixed and then applied to a subcarrier amplifier. The mixed pulse controls operation of the subcarrier amplifier so that the subcarrier is not generated during the blanking period, except for the burst flag period. The subcarrier amplifier generates both the inverted and noninverted subcarrier signals to balance-modulate the R-Y- and B-Y-signals. The two balance-

Sec. 4-6   Camera Signal-Processing Circuits (with MOS pickup)

modulated color difference signals are mixed to generate the chroma signal in the mixer.

The mixed chroma signal is applied to a fade/gain-control circuit where gain of the output signal is controlled by input level. If input level drops, gain is increased, and vice versa. This maintains the chroma output at a constant level. RM1102 sets the fade/gain-control level.

The chroma signal is applied to the Y/C mixer through a noise-reducer circuit and a 3.58-MHz BPF. The noise reducer reduces random noise by adding together the even and odd horizontal line (1H) signals. The 1H-delayed chroma signal is added to the chroma signal that appeared 1H before and produces an average chroma signal with reduced noise. The 3.58-MHz BPF passes only the 3.58-MHz chroma signals to the Y/C mixer.

4-6.12 Date/Time Character Generator Circuits

Figure 4-46 shows the date/time character generator circuits used in a typical camcorder with MOS sensor.

Most of these circuits are contained within generator IC1402. When 5-V power is applied to IC1402 at pin 15, crystal X1401 generates a 32.768-kHz clock pulse, and the time microprocessor in IC1402 starts operation. CT1401 provides a means to adjust the clock frequency.

When power is turned off, or the operating mode is changed from camera to VCR (or VTR on some camcorders), the 5-V power is removed from pin 15 of IC1402. At this time, a backup power supply (DC 3 V) is applied through D1403 and pin 8 of IC1402. This keeps IC1402 operating to retain date/time information.

The character generator in IC1402 receives display data information (date/time) from the timer and generates corresponding character signals synchronized with the $\overline{CHD}$ and $\overline{FV}$ signals from the sync generator IC1302 (Sec. 4-6.2). The character signal is applied to the base of Q1303. When the character signal is high, Q1303 turns on, and the character is added to the video output from encoder IC1301 (Sec. 4-6.11).

The display mode for date and time is selected by the level at pins 3, 5, 24, 25, and 26 of IC1402, as shown by the tables in Fig. 4-46. In the configuration shown, the date display is set to the month, day, year, format, since pin 25 is high (5 V), while time is set to the 12:15 A.M., 11:15 P.M. format, since pins 3, 5, 24, and 26 are low (ground or 0 V).

The date/time setting and display switches S1607 to S1610 determine the display generated by the character generator.

When display switch S1607 is pressed, the character display appears in the EVF screen. The DISPLAY mode is changed to date, time, date/time, and display off, repeatedly, every time S1607 is pressed.

When shift switch S1609 is pressed, the display blinks. The blinking

Date Display Mode

Pin		Display mode
25	24	
0	0	Day, month, year
0/1	1	Year, month, day
1	0	Month, day, year

Time Display Mode

Pin			Display mode (Example = 12: 15 AM, 11:15 PM)
5	3	26	
0	0	0	12:15 AM 11:15 PM
0	1	0	0:15 AM 11:15PM
0	1	1	AM 0:15 PM 11:15
1	0	1	AM 12:15 PM 11:15
1	1	1	0:15 23:15

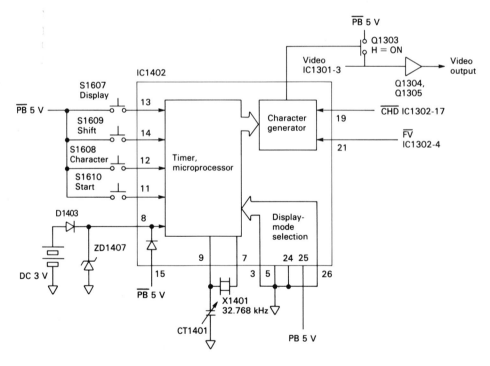

FIGURE 4-46 Date/time character generator circuits for camcorder with MOS sensor

Sec. 4-6   Camera Signal-Processing Circuits (with MOS pickup)   139

display is changed to A.M. (P.M.), hour, minute, year, month and day, repeatedly, every time S1609 is pressed.

When character switch S1608 is pressed, the blinking number is incremented. In case of the A.M. or P.M. blinking display, A.M. is changed to P.M. alternately every time S1608 is pressed.

Start switch S1610 is used to start and stop the setting for date and time. When S1610 is first pressed for setting date or time, the A.M. or P.M. display blinks, and the setting can be made. When S1610 is pressed after setting the date or time, blinking stops and date/time counting starts.

4-6.13   MOS sensor signal-processing circuit troubleshooting

The basic troubleshooting approach described in Sec. 4-4.12 also applies to the signal-processing circuits of camcorders with MOS sensors. Of course, there are obvious differences. Compare Figs. 4-16 and 4-33.

When tracing the video signal from the MOS sensor to the chroma encoder (Fig. 4-33), pay particular attention to the following test points. Check the service literature for correct values at corresponding points on the camcorder you are servicing.

The output of the MOS sensor can be monitored at pins 17–20 of IC1001. Typically, these outputs are in the 40- to 50-mV p-p range. The outputs should increase when the camera is aimed at a color chart. This generally applies to all video signal levels in the path from the MOS sensor to the NTSC encoder.

If there is no output from IC1001, check that IC1001 is receiving drive signals from IC1006. The drive signals should be about 5.0 V p-p.

The output of the video preamps can be monitored at pin 8 of IC1002-IC1005. These outputs to IC1101 should be about 4.0 V p-p.

The outputs of the matrixing circuits can be monitored at pins 2, 4, 5, 6, 7, 9, and 10 of IC1101. The Y-signal output at pin 2 should be about 1.35 V p-p. This signal drops through DL1100 to about 1.3 V p-p at pin 1 of IC1102. The color signal outputs from IC1101 should be (approximately) as follows: -G pin 4 = 0.8 V p-p, +G pin 5 = 4.0 V p-p, -R pin 6 = 3.6 V p-p, +R pin 7 = 0.7 V p-p, -B pin 9 = 2.2 V p-p, +B pin 10 = 2.8 V p-p.

The outputs of the filters can be monitored at pin 5 of IC1102 and pin 8 of IC1103-IC1105 (or at TP1102-TP1104). The Y-signal at pin 5 of IC1102 should be about 3.9 V p-p. The color signal at pin 8 of IC1103-IC1105 should be about 1.2 or 1.3 V p-p.

The color signal outputs from processor IC1107 should be about 0.65 V p-p at pin 16 and 0.15 V p-p at pin 17, while the Y-signal output at pin 14 should be about 0.9 V p-p. The AIC control output at pin 15 of IC1107 should be about 0.44 V p-p.

If there are good signals at pins 13–15 of IC1301, but you do not get a good NTSC signal (video output), suspect IC1301 (Sec. 4-6.11).

If there is good video output, but no date/time display, suspect IC1402 and Q1303 (Sec. 4-6.12).

If the iris appears to be inoperative, suspect the iris motor and IC1108 (Sec. 4-6.8).

If there appears to be a problem with white balance, set S1601 to MANUAL and check if good white balance can be obtained manually by adjustment of RT1602. Then set S1601 to AUTO, and check automatic operation of the white-balance circuits. If you get good white-balance manually with RT1602, but not automatically, suspect IC1403 (Sec. 4-6.10). If you do not get good white balance manually or automatically, suspect IC1107 (Sec. 4-6.9).

Keep in mind that all the circuits shown in Fig. 4-33 depend on pulses from sync generator IC1302 (Sec. 4-5.2).

5

TAPE TRANSPORT and SERVO SYSTEM

This chapter describes the theory of operation for typical camcorder tape-transport and servo system circuits, as well as operation of mechanical components. All the notes described in the introduction of Chapter 4 apply here.

5-1 TAPE TRANSPORT BASICS

The tape-transport mechanism of a VHS camcorder is very similar to that of a VHS VCR. The camcorder mechanism uses standard-sized VHS VCR, so the dimensions for the cassette compartment are the same. To maintain compatibility between tapes recorded on the camcorder and other VHS VCRs, the tape path is essentially the same. As a result, the mechanical components (guideposts, angle posts, full-erase head, etc.) are located in about the same areas.

As discussed in Chapter 1, the diameter of the cylinder is reduced to make the tape transport smaller. However, to maintain compatibility with conventional VCRs, additional variations are necessary to compensate for the reduction in cylinder diameter.

In a conventional VHS VCR, the videotape is wrapped approximately half way (M-load) around the cylinder (180°). In a VHS camcorder, the videotape wrap is approximately three quarters of the way around the upper cylinder assembly (270°). This is necessary to get the same length of tape around the surface of the heads on the cylinder. Because the cylinder is smaller in diameter, cylinder rotation speed is increased to 2700 rpm in a camcorder, compared to 1800 rpm for a conventional VHS VCR.

Another major difference is that a VHS camcorder uses four video heads to duplicate the performance of a standard 2-head VHS VCR. Because the cylinder is smaller and cylinder rotation is faster, four heads are necessary to record or play back the video tracks correctly.

In a conventional 2-head VCR, one cylinder revolution records two video tracks (one field per head) which is one video frame. A VHS camcorder maintains the same characteristic or recording or playing back one field per head. Since the cylinder is smaller and rotates faster, the cylinder records one field in three quarters of a revolution, or one frame (two fields) per $1\frac{1}{2}$ revolutions. The resulting tracks duplicate the path left by a standard 2-head VHS VCR during an SP recording.

To prevent adjacent track pickup during playback, each video head is turned on when scanning a specific field. This requires a very sophisticated headswitching scheme. We discuss head switching in this chapter, and in Chapter 6.

5-2 MECHANICAL OPERATION OF TYPICAL VHS CAMCORDER

This section describes operation for the mechanical parts (tape transport, loading mechanism, brake, etc.) of a typical VHS camcorder. Since operation of the mechanical parts are closely related to the servo system, it is essential that you review the servo circuit descriptions found in remaining sections of this chapter. Likewise, since operation of both the mechanical parts and servo system are under direct control of the system microprocessor, you must also study Chapter 7. Note that Chapter 8 describes adjustment procedures for the mechanical parts, as well as electrical adjustments.

By studying the mechanical operation found here, you should have no difficulty in understanding the mechanical operations of similar VHS camcorders. This understanding is essential for logical troubleshooting, no matter what type of mechanical equipment is involved. For example, if you know that a particular solenoid is actuated to pull a certain rod in a given mode of operation, and you see that the solenoid does not actuate in the given mode, you have pinpointed a failure. The origin of the trouble may be electronic (no actuating signal is present) or mechanical (the solenoid is jammed), but you have a starting point for troubleshooting. The descriptions given here should also help you interpret the mechanical sections of VHS camcorder service literature (which are often vague, or simply omitted).

Sec. 5-2   Mechanical Operation of Typical VHS Camcorder   143

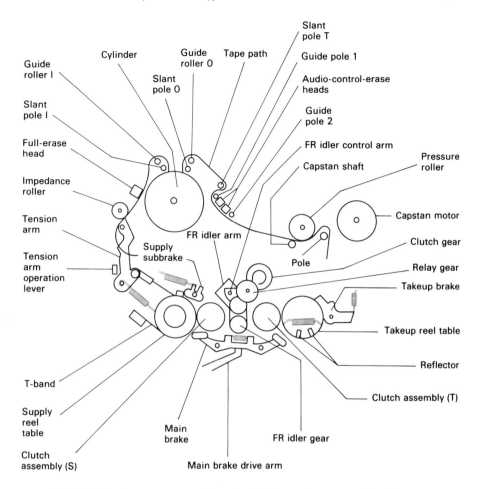

FIGURE 5-1 Arrangement of mechanical parts from top side of tape transport

5-2.1 Arrangement of Mechanical Parts

Figure 5-1 shows arrangement of the mechanical parts from the top side of the tape transport, while Figure 5-2 shows the arrangement from the bottom. Figure 5-3 shows a side view of the mechanism. Note that the configuration shown is for a specific VHS camcorder, but it will not be substantially different for most VHS camcorders.

Since the recording tape pattern is the same as the VHS standard, the positions and arrangement of basic parts are the same. However, the loading

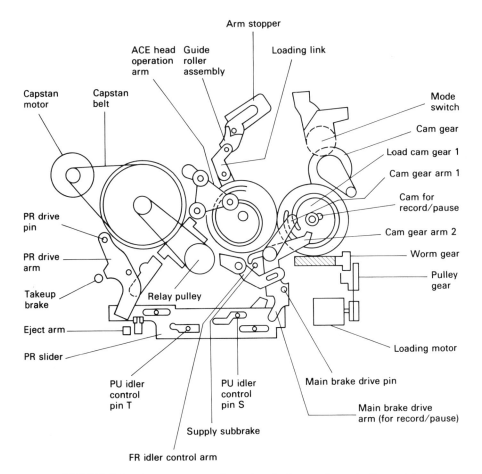

FIGURE 5-2 Arrangement of mechanical parts from bottom side of tape transport

mechanism uses a modified M-loading system, since the tape must be wrapped around the cylinder (for about 270°).

The *impedance roller* is essentially the same as for most VHS VCRs and must be adjusted in height (Chapter 8).

Since loading is modified M-loading, the *ACE head* moves out of the way during loading. After loading is complete, the ACE head rotates and sets itself at the prescribed position.

The *reel tables* are the same as for a typical VHS VCR. The takeup reel has a *reflector for rotation detection* (Chapter 7).

The *supply* and *takeup clutch assemblies* are also used as a torque limiter for the reel tables. During fast forward and rewind, a direct coupling disengages the torque limiter from the reel tables and drives the clutch. During

Sec. 5-2 Mechanical Operation of Typical VHS Camcorder   145

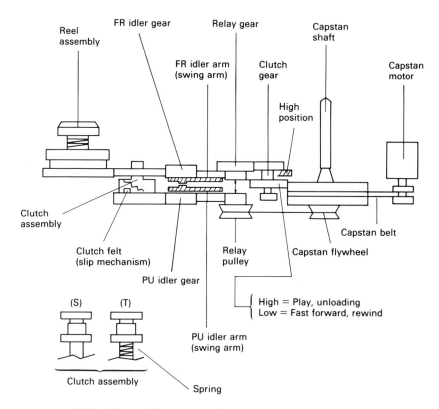

FIGURE 5-3 Side view of tape-transport mechanism

playback (take up) and unloading, the clutch is driven by weak torque coming through the limiter.

The *FR driver* is the swing mechanism found in many VHS VCRs and is operated only during fast forward and rewind.

The *PU idler* is attached to the bottom of the FR idler and engages at the bottom of the clutch assembly during unloading (when weak torque is required) and during playback (when slip is required). The PU idler uses the clutch assembly slip mechanism to limit torque while driving both reel tables.

The *relay pulley* transmits torque to both real tables from the *capstan flywheel* and is a clutch gear of the type used in many VHS VCRs. Since the gear assembly disengages when the gear on the bottom is pushed upward, torque transmitted to both reel tables passes through the clutch assembly friction mechanism and, thus, transmits a weak force. When the bottom gear in the clutch gear assembly is in the low position (Fig. 5-3), weak force is transmitted to both reel tables by direct gear contact.

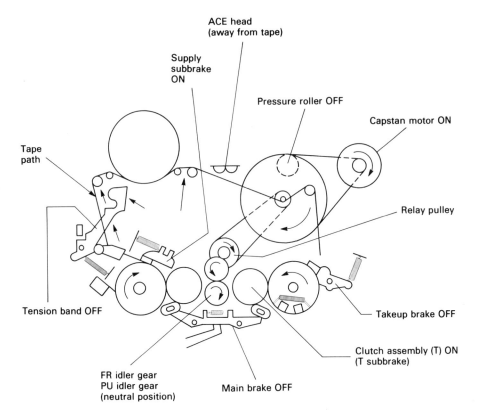

FIGURE 5-4 Relationship of mechanical parts and tape path in loading mode

5-2.2 Operation of Mechanical Parts During Loading Mode

Figure 5-4 shows the relationship of the mechanical parts, as well as the tape path, during the loading mode. Also refer to Figs. 5-1 through 5-3.

Torque from the loading motor is transmitted through the pulley belt to the pulley gear and is then transmitted through the worm gear to both load cam gear I (input) and load cam gear O (output). A loading link mechanism attached to the load cam gear I, and O pulls out the tape from the cassette as the loading motor rotates.

After the tape has been fed out, the ACE head operation arm sets the ACE head into the correct position. At the same time, the PR slider slides and turns the PR drive arm by rotating the main brake drive arm, and the pressure roller is moved against the capstan shaft.

During loading, the capstan motor rotates clockwise (Fig. 5-4). This

Sec. 5-2   Mechanical Operation of Typical VHS Camcorder   147

clockwise rotation is transmitted through the relay pulley to the PU idler and keeps the PU idler at the center (or neutral) position.

5-2.3 Operation of Mechanical Parts During Unloading Mode

Figure 5-5 shows the relationship of the mechanical parts, as well as the tape path, during the unloading mode. Also refer to Figs. 5-1 through 5-3.

Pressing the stop button in the playback mode operates loading and, in reverse sequence, unloading. During unloading, the PU idler is pushed

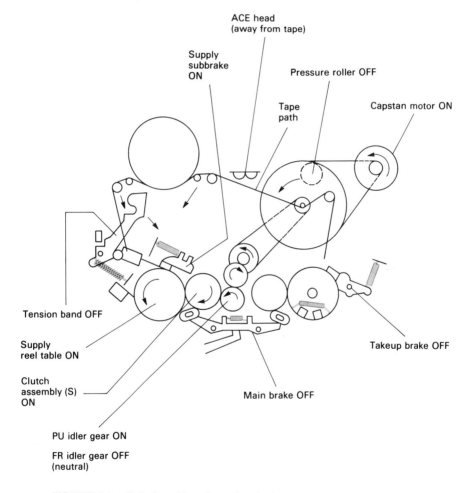

FIGURE 5-5 Relationship of mechanical parts and tape path in unloading mode

toward clutch assembly (S) to transmit limited torque to the supply reel table and thus permit unloading with the proper tension. When tape wind is complete, the main brake operates and places the mechanism in stop.

5-2.4 Operation of Mechanical Parts During Playback/Record Mode

Figure 5-6 shows the relationship of the mechanical parts, as well as the tape path, during normal forward operation (playback/record). Also refer to Figs. 5-1 through 5-3.

The tension-control mechanism (T-band) controls tension of the tape fed from the supply reel at a fixed rate. The flange on the lower edge of the impedance roller, which sends the tape to the cylinder, regulates the height of the tape.

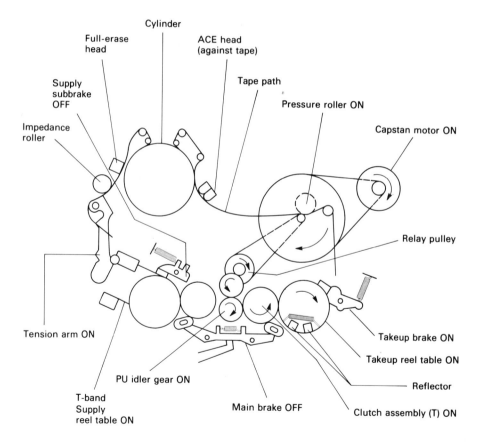

FIGURE 5-6 Relationship of mechanical parts and tape path in normal forward mode (playback/record)

Sec. 5-2 Mechanical Operation of Typical VHS Camcorder **149**

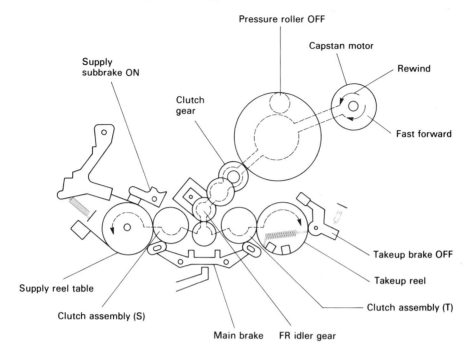

FIGURE 5-7 Relationship of mechanical parts and tape path in both fast-forward and rewind modes

The tape passes over the cylinder and is taken up by the takeup reel. At this time, the clutch assembly (T) slip mechanism slips and takes up the tape.

An infrared sensor detects rotation of the reflector attached to the bottom of the takeup reel and produces pulses as the takeup reel rotates (as described in Chapter 7).

5-2.5 Operation of Mechanical Parts During Fast Forward/Rewind Modes

Figure 5-7 shows the relationship of the mechanical parts, as well as the tape path, during both fast forward and rewind. Also refer to Figs. 5-1 through 5-3.

During fast forward, the PR idler is pushed into contact with clutch assembly (T) to take up the tape rapidly. The supply subbrake is also pushed into contact with the supply reel table, and applies back tension so there will be no slack in the tape. Since the clutch gear is in the bottom position (Fig. 5-3), capstan motor torque is transmitted to the takeup reel with no slip.

Rewind is the reverse of fast forward, and is accomplished by rotating the capstan motor in reverse. Since the takeup subbrake is under clutch assembly (T), the spring (Fig. 5-3) rotation loss provides the proper back tension.

5-2.6 Operation of the Brake Mechanisms

Figure 5-8 shows the relationship of the mechanical parts associated with the brake mechanisms. Also refer to Figs. 5-1 through 5-3.

When the mechanism is stopped and the playback button is pressed, the loading motor starts to rotate in the loading direction. The main brake drive pin (Figs. 5-2 and 5-8) moves to the right. The brake slider (Fig. 5-8) connected to the pin also moves to the right. The protrusion on the brake slider locks into the holes in the lock arm and moves the lock arm to the right. The beveled section main brake operation slider, connected by the

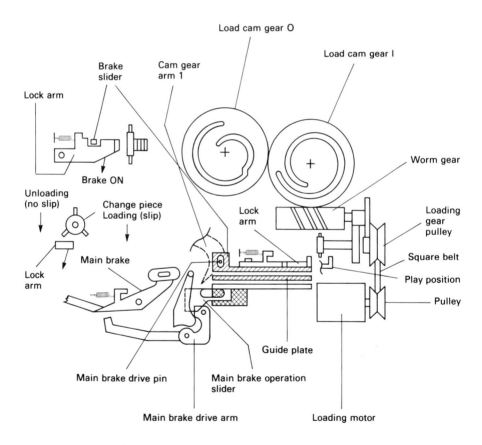

FIGURE 5-8 Relationship of mechanical parts associated with the brake mechanism

lock arm and pin, rotates the main brake drive arm and disengages the main brake.

When the mechanism is in the normal FORWARD mode (playback/record) and the stop button is pressed, the loading motor rotates in the unloading direction and causes the tape to be taken up on the supply reel.

This operation also starts the return of the lock arm to the left. When the lock arm contacts any of the three protrusions on the change piece, the lock arm is moved upward. This disconnects the lock arm and brake slider. The lock arm then moves quickly to the left and engages the main brake.

Keep in mind that the brake system provides a means to stop tape travel mechanically. The capstan and loading motors must also be stopped electrically by the servo system (described in the following paragraphs) and by the system control microprocessor (described in Chapter 7).

5-3 SERVO SYSTEM BASICS

The cylinder and capstan servo systems found in VHS camcorders are very similar to those of VHS VCRs. Typically, camcorder servo systems use a PWM (pulse-width modulation) phase-loop and an analog (sample-and-hold) speed-loop configuration.

Figure 5-9 shows a typical camcorder servo system in block form, together with a summary of the servo system signals. Readers familiar with the VCR servo systems will recognize most of the circuit elements, with the possible exception of IC610. This integrated circuit produces the necessary pulses to accommodate the 4-head cylinder that rotates at 2700 rpm.

When the cylinder rotates at 2700 rpm (instead of the usual 1800 rpm), a 45-Hz cylinder tach pulse is developed (instead of the usual 30-Hz pulse). IC610 converts the 45-Hz pulse from the cylinder into a 30-Hz pulse that is applied to IC601. (The circuits within servo IC601 are almost identical to those found in VCR servo ICs.)

The 30-Hz pulse (tach pulse) from IC610 is applied to IC601 and is returned back to IC610 as the 30-Hz switching pulse (SW 30 Hz), together with a 360-Hz cylinder FG pulse. These two signals are used to generate the proper head-switching signals for the four video heads (as discussed in Sec. 5-5 and Chapter 6).

The servo system shown in Fig. 5-9 also includes a lock-search circuit IC608 during search operation. Lock-search is achieved by dividing down both the capstan FG and the control track pulse (CTL pulse) by a factor of 3 (from 2160 to 720 Hz) in the search mode (either forward or reverse). This maintains normal operation of the capstan servo system during search operation.

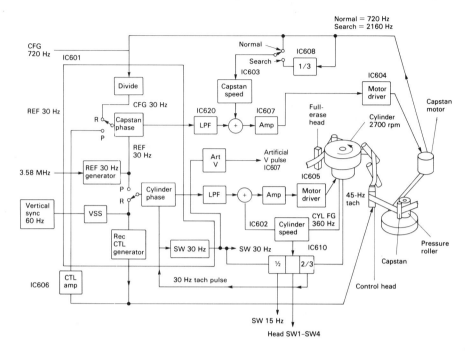

Motor	System	Mode	Reference Signal	Comparison signal
Cylinder	Phase	Record	1/2-V SYNC	30-HZ PG tach pulse
		Play	REF 30 Hz	
	Speed	Shared	Cylinder FG (CYL FG: 360 Hz)	
Capstan	Phase	Record	REF 30 Hz	Capstan FG (CFG: 720 Hz)
		Play		CTL pulse (30 Hz)
	Speed	Shared	Capstan FG (CFG: 720 Hz)	

FIGURE 5-9 Typical camcorder servo system

5-3.1 Servo Circuit Operation

Figure 5-10 shows the basic servo circuit configurations in block form. These servos maintain tape speed (capstan servo) and video head rotation speed (cylinder servo), at a constant value.

During record, the magnetic tape must run at a fixed speed of 3.335 cm/sec. The video heads must rotate at 2700 rpm to produce a 45-Hz signal

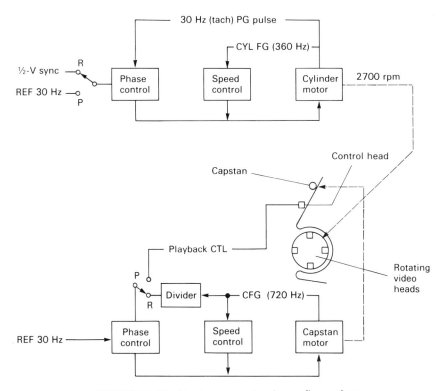

FIGURE 5-10 Basic servo circuit configuration

(which is three quarters of the video vertical sync signal frequency). Also, the relative position of the heads and tape must be maintained so that the control head can record the vertical blanking interval at the bottom edge of each track.

During play, the tape must maintain the same running speed as in record, the video heads must rotate at 2700 rpm (45 Hz), and the heads must trace the recorded video track accurately.

To accomplish these objectives, both the phase and speed of the capstan and cylinder motors must be controlled. The phase-control system controls small fluctuations while the speed-control system controls large fluctuations. When the effects of the phase and speed systems are combined the capstan motor runs at the rated tape speed, and the cylinder motor rotates at a fixed speed. The combinations of phase and speed control suppress wow and flutter, particularly during start-up (an essential feature for camera operation).

5-3.2 Origin and Purpose of Servo System Signals

The $\frac{1}{2}$-*V sync signal* is obtained by dividing the video signal vertical sync frequency (60 Hz) in half. The $\frac{1}{2}$-V sync signal phase is compared with that of the 30-Hz PG pulse to control cylinder motor phase during record.

The REF 30-Hz signal is obtained by dividing the 3.58-MHz chroma subcarrier signal (by 119318). The REF 30-Hz signal is used as a reference signal to control phase of the cylinder motor during playback. The capstan motor-control system also uses the REF 30-Hz signal as a reference for phase control, during both record and play.

The 30-Hz PG pulse (also called a tach pulse) is produced by the cylinder and IC610. A disk is attached to the bottom of the cylinder motor shaft. A magnet is mounted on the disk. As the magnet rotates with the cylinder motor, the magnet passes a Hall device during each revolution. The Hall device produces a 45-Hz pulse which is converted to a 30-Hz pulse by IC610. The 30-Hz PG pulse is used to detect video head position and cylinder speed. The phase of the cylinder motor is controlled by comparing the phase of the 30-Hz PG pulse with that of the $\frac{1}{2}$-V sync pulse.

The cylinder FG (CYL FG) pulse is a 360-Hz pulse generated by the stator coil and 16-pole magnet attached to the rotor on the cylinder motor. the cylinder FG pulse controls the speed of the cylinder motor during both record and play operations.

The capstan FG (CFG) pulse is a 720-Hz pulse generated by an FG board and 48-pole magnet attached to the rotor on the capstan motor. The capstan FG pulse and the REF 30-Hz signal phases are compared during record to control the phase of the capstan motor. The capstan FG pulse also controls the speed of the capstan motor during both record and play.

The CTL (or control) pulses are derived from the incoming vertical sync during record and are applied to the tape by the A/C (audio/control) head. During playback, the phases of the recorded CTL pulse and the REF 30-Hz signal are compared to control capstan motor phase.

5-4 CYLINDER CONTROL

Figure 5-11 shows the cylinder servo circuit in block form. IC602 controls speed while IC601 controls phase of the cylinder motor. Output signals from IC601 and IC602 are added and then applied to the cylinder motor through drive circuits (IC607 and IC605).

Speed-control IC602 samples the 360-Hz cylinder FG pulse and produces the corresponding speed-control output at pin 15.

Phase-control IC601 samples the 30-Hz PG pulse (developed by IC610) and compares the pulse with either the $\frac{1}{2}$-V sync pulse (during record) or the REF 30-Hz signal (during playback) to produce the corresponding phase-control output at pin 29. This output is applied through LPF IC620 to the adder circuit.

5-4.1 Cylinder Speed Control

Figure 5-12 shows typical cylinder speed-control circuits.

The FG amplifier in IC602 amplifies the cylinder FG pulse (360 Hz)

Sec. 5-4 Cylinder Control   155

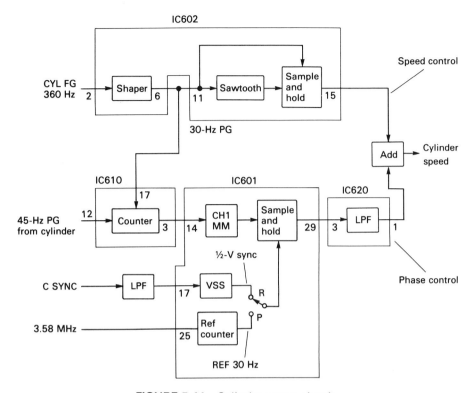

FIGURE 5-11 Cylinder servo circuit

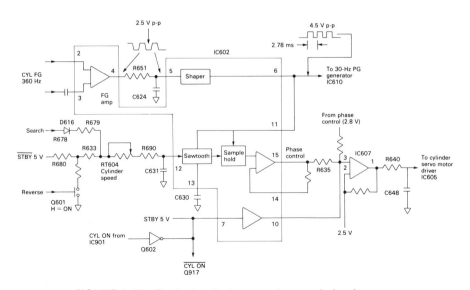

FIGURE 5-12 Typical cylinder speed-control circuits

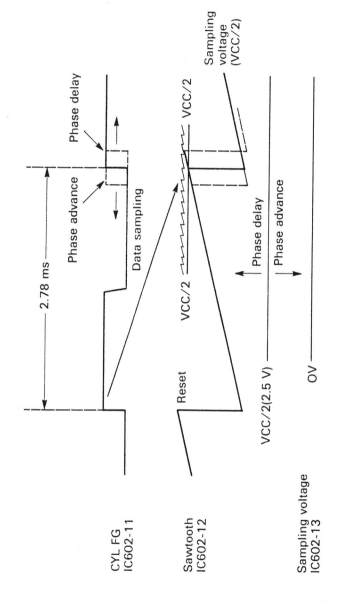

FIGURE 5-13 Cylinder speed-control timing

Sec. 5-4   Cylinder Control   157

from the cylinder motor and applies the pulse through an LPF (C624/R651) to the shaper. The shaped cylinder FG pulse is output at pin 6 of IC602 and is applied to IC610. The output at pin 6 of IC602 is also returned to sample-and-hold circuits within IC602.

A sawtooth generator produces a sawtooth signal for each period from the leading edge of the cylinder FG pulse, as shown in Fig. 5-13. The time-constant circuit (R678, R633, R690, and C631, and cylinder speed-control RT604) determines the level of the sawtooth signal, as adjusted by RT604.

The sawtooth signal is applied to the sample-and-hold circuit and is sampled by the leading edge of the previous cylinder FG pulse, as shown in Fig. 5-13. The sampling information is held until C630 (at pin 13 of IC602) detects the sampling information from the next period.

The rated sampling voltage is Vcc/2, or about 2.5 V. As the load changes (causing cylinder speed to change), the cylinder FG pulse period changes, causing the sawtooth sampling position to change. As the load increases, the sampling voltage goes higher than 2.5 V, and the output level rises. As the load decreases, the sampling voltage becomes less than 2.5 V, and the output level decreases.

The signal output from the sample-and-hold circuit is amplified and appears as a d-c control voltage at pin 15 of IC602. This signal is added to the cylinder phase-control signal (Sec. 5-4.5), to create the cylinder servo signal that is applied through a filter to the motor driver IC605.

5-4.2 Cylinder Motor Inhibit

Figure 5-12 shows the cylinder motor inhibit circuits. During stop, fast forward, and rewind modes, the cylinder motor is inhibited (so that the cylinder will not grind up the tape). This inhibit operation in controlled by the system-control microprocessor IC901 (Chapter 7). The inhibit signal (CYL ON) from IC901 is applied to Q602.

During stop, fast forward, and rewind, the CYL ON signal from IC901 goes low. This turns Q602 off, causing the input to inverter at pin 7 of IC602 to go high. The inverter output goes low, grounding pin 10 of IC602, and shorts the control output to the cylinder motor.

5-4.3 Cylinder Motor Visual Search Operation

Figure 5-12 shows the cylinder motor visual search-control circuits. In the visual SEARCH modes (forward or reverse), the cylinder motor phase-control system is inhibited, since the cylinder speed is changed from normal. In forward search, the cylinder motor is set at 2730 rpm (instead of 2700). In reverse search, the cylinder motor is set at 2641 rpm. These operations are controlled by the system-control microprocessor IC901 (Chapter 7). IC901 supplies a search signal to D616 and a reverse signal to Q601.

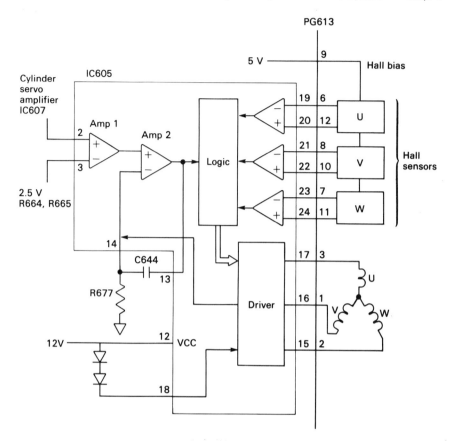

FIGURE 5-14 Typical cylinder motor-drive circuits

When either forward or reverse search operation is selected, IC901 applies a high to D616. This turns D616 on and connects R679 into the circuit. If forward search is selected, the cylinder motor is speeded up to 2730 rpm. If reverse search is selected, IC901 applies a high to Q601. This turns Q601 on and connects R680 into the circuit. The cylinder motor is slowed down to 2641 rpm.

5-4.4 Cylinder Motor Drive

Figure 5-14 shows typical cylinder motor-drive circuits.

The cylinder motor control signal from IC607 is applied to pin 2 of the cylinder motor driver IC605. The control signal is amplified by amp 1 and amp 2 and applied to the logic circuit. The voltage drop across R677 determines the gain of amp 2. This voltage drop is also used to determine the

Sec. 5-4   Cylinder Control

load on the cylinder motor. If load increases to slow cylinder rotation, the gain is altered to increase rotation, and vice versa.

The logic circuit receives rotor position data from the cylinder motor phase-position sensors (Hall sensors) and applies drive signal data (determined by rotor position) to the driver. In turn, the driver applies drive signals to the cylinder motor windings.

5-4.5 Cylinder Speed-Loop Troubleshooting

If the picture is out of horizontal sync on both the EVF and TV (the most common symptom for trouble in the cylinder phase/speed circuits), play a known-good tape or a test tape (Chapter 3). Then try correcting the condition by adjustment of the cylinder speed control RT604 (Fig. 5-12).

As discussed in Chapter 8, the first step is adjusting RT604 is to disable the phase-control circuits (so that the phase loop will not try to lock up and correct a malfunction in the speed loop). You then adjust RT604 until a sampling pulse is stationary (or moves slowly) on the PG pulse. The horizontal sync should also be restored. (However, you may get noise bars floating in the picture with the phase-loop disabled.)

As a next step, you can try correcting the problem by troubleshooting the phase loop as described in Sec. 5-4.12. However, if the cylinder motor speed cannot be locked by adjustment of RT604, suspect IC602 and the associated circuits. Proceed as follows.

First, check for 360-Hz cylinder FG pulses at pin 6 of IC602. If missing, suspect a defective FG generator in the cylinder motor assembly, IC602, or the low-pass filter (C624/R651).

Next, check for a d-c signal of about 2.5 V at pin 15 of IC602. If missing, suspect a defect with the sawtooth generator network connected at pin 12 of IC602, or the sample-and-hold capacitor C630 at pin 13, or a circuit within IC602.

Next, check for a d-c signal of about 2.8 V at pin 1 of IC607. If absent or abnormal, suspect IC607 or a missing reference voltage at pin 2 of IC607.

If the normal voltage is obtained at pin 1 of IC607, the cylinder speed-loop is operating correctly. Under these conditions, if the cylinder motor is not rotating at the correct speed, the problem is likely in IC605 or the cylinder motor itself (Fig. 5-14).

5-4.6 Cylinder Phase Control

Figure 5-15 shows typical cylinder phase-control circuits.

After the speed-control system sets the speed of the cylinder motor to about 2700 rpm, the cylinder motor phase control adjusts the video signal vertical retrace period so that the period is recorded on the trace start point (bottom edge of tape) of each track. During play, cylinder motor phase-

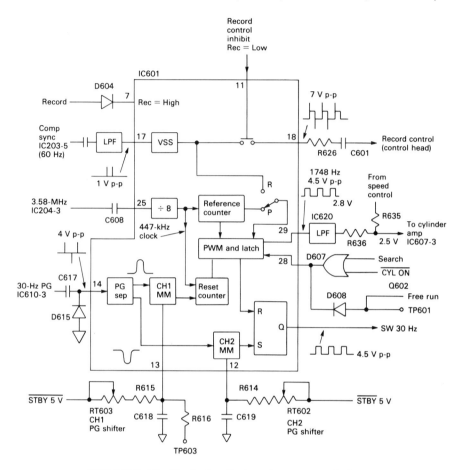

FIGURE 5-15 Typical cylinder phase-control circuits

control circuits control the video heads so that they are always positioned on the recorded vertical sync signal.

During record, the 30-Hz $\frac{1}{2}$-V sync signal (taken from the camera or other video signal) is used as one reference. During play, the REF 30-Hz signal (taken from the 3.58-MHz chroma subcarrier) is used as the reference.

During either record or play, the 30-Hz PG or tach pulse (taken from the 45-Hz cylinder pulse through IC610) is used as the second reference.

5-4.7 30-Hz PG or Tach Pulse Generator

Figure 5-16 shows typical 30-Hz PG or tach pulse generator circuits.

To use the same servo circuit configuration for control found on conventional VCRs, the 360-Hz cylinder FG pulse is combined in IC610 with the 45-Hz cylinder tach pulse to produce a 30-Hz PG or tach pulse.

Sec. 5-4   Cylinder Control   161

The 45-Hz signal is applied to a monostable multivibrator (MM) through an amplifier in IC610. The multivibrator has a fixed time constant (determined by C647/R687) that increases the pulse width of the 45-Hz signal. This is done so that the falling edge of the multivibrator output signal occurs after the rising edge of the corresponding cylinder FG pulse (as shown in the timing chart of Fig. 5-16).

The multivibrator output is divided down to a 15-Hz reset pulse and supplied to a divide-by-12 counter. The divide-by-12 counter divides the 360-Hz cylinder FG pulses, but is reset every 8-cylinder FG pulses by the 15-Hz divided-down cylinder FG pulse. The result is a 30-Hz PG or tach pulse (used by cylinder phase control and the head-switching circuits, Sec. 5-9).

As shown in the timing chart of Fig. 5-16, the 45-Hz PG signal (1 pulse

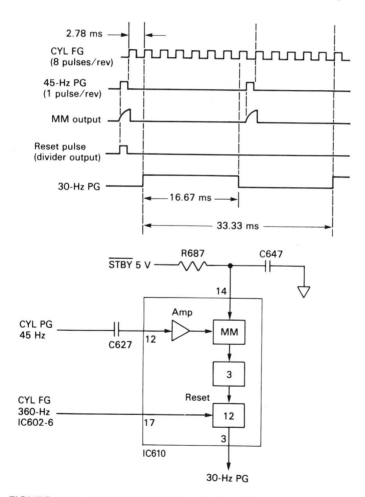

FIGURE 5-16 Typical 30-Hz PG or tach pulse generator circuits

per revolution) of the rotating cylinder is delayed by the multivibrator to ensure that the division sequence starts at a complete cycle of a cylinder FG pulse. The divided-down (divide by 3) 45-Hz PG pulse becomes the reset pulse for the divide-by-12 counter. As a result, each 24-cylinder FG pulses are divided down to two 30-Hz PG or tach pulse signals.

5-4.8 Comparator Signal Input Circuit

Figure 5-15 shows the comparator signal input circuit within IC601.

The 30-Hz PG or tach pulse is applied to a tach separator that produces two pulses (of opposite polarity). These pulses are applied to multivibrators that electrically correct any mechanical distortion in the video head and tach magnet positions (as described in Sec. 5-4.11). The output of the channel 1 multivibrator is applied to a counter. The output of the channel 2 multivibrator is applied to an RS flip-flop.

5-4.9 Reference Signal Input

Figure 5-15 shows the reference signal input circuits.

During record, the low-pass filter passes a vertical sync signal (60 Hz) from the video signal input. The vertical sync signal is applied to a vertical sync signal (VSS) divider and results in a 30-Hz $\frac{1}{2}$-V sync signal. The 30-Hz signal then passes through the record/playback switch to a latch circuit as a latch pulse.

During play, the 3.58-MHz carrier signal in applied to a $\frac{1}{8}$-frequency divider where a 447-kHz clock pulse is obtained. The clock pulse is then applied to a reference counter and to a reset counter. The reference counter divides the frequency of the 447-kHz pulse into a REF 30-Hz signal, which is applied to the latch circuit through the record/playback switch.

5-4.10 Sample and Hold

Figure 5-15 shows the sample-and-hold circuit. Figure 5-17 shows the corresponding waveforms.

The sample-and-hold circuit consists of a counter and latch. The counter receives PG or tach pulses at the reset input from multivibrator CH1 and counts the 447-kHz clock pulses. The counter outputs data to the latch where the rise in the latch pulse (reference signal) holds the data in the latch (as a comparison signal).

The phase of the reference signal and comparison signal are compared to produce a difference signal that is applied to the PWM. The difference signal modulates the duty cycle of the 1748-Hz signals from the reference counter. The signal from the PWM is converted to a d-c voltage by the LPF within IC620.

Sec. 5-4 Cylinder Control   **163**

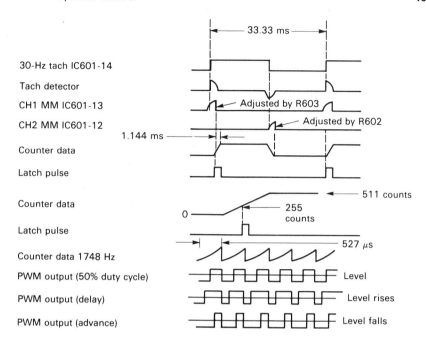

FIGURE 5-17 Cylinder phase-control/PWM output timing

Note that the reset counter overflows at a count of 511. However, during normal operation, the count is 255 to produce a PWM output with a 50% duty cycle. When the load increases (cylinder starts to slow down), the counter produces more than 255 pulses, indicating a phase delay. The PWM duty cycle then goes above 50%, and the output level (smoothing level) rises. The opposite occurs when the load decreases (cylinder starts to speed up).

The d-c voltage from IC620 is added to the control voltage from the speed-control circuit. The sum of the two voltages is applied to the cylinder motor through cylinder amplifier IC607 and cylinder motor drive IC605.

During record, the vertical sync signal at pin 17 of IC601 is also applied through a control record-inhibit switch to pin 18 of IC601. These 30-Hz pulses drive the control track record head to place accurate timing marks on the bottom edge of the video tape (in proper relation to the vertical sync of the incoming signal). As in all VCRs, the control track is used during playback to maintain proper speed of the capstan motor (Sec. 5-5).

An inhibit signal is applied to pin 28 of IC601 to stop operation of the cylinder phase loop. A high at pin 28 turns off the PWM system and results in a fixed 50% duty-cycle output at pin 29. This is done during search mode, as described in Sec. 5-4.3, by the search signal applied to pin 28 of IC601 through D607.

As discussed in Sec. 5-4.2, the cylinder motor is turned off during stop, fast forward, and rewind. Under these conditions, the PWM output is locked to a fixed 50% duty cycle by a $\overline{\text{CYL ON}}$ signal from Q602 applied to pin 28 of IC601 through D607.

Pin 28 of IC601 also serves as an input for a *free run test signal* applied through D608. As discussed in Chapter 8, the free-run test point is used during setup and adjustment of the cylinder speed-loop circuit. By connecting the free-run test point to +5 V, a high is applied to pin 28 of IC601 and the PWM circuit produces a fixed 50% duty-cycle output. This is done to eliminate the phase control from affecting the control voltage (from pin 29) to the cylinder motor and permits correct setting of the cylinder speed control.

5-4.11 SW 30-Hz Signal Generator

Figure 5-15 shows the SW 30-Hz signal generator circuit. Figure 5-18 shows the corresponding waveforms.

The channel 1 (CH1) and channel 2 (CH2) multivibrators have a variable time-constant network connected to pins 13 and 12 of IC601, respectively. These variable timing networks are referred to as the channel 1 and channel 2 PG shifter networks (in most camcorder literature).

The shifter adjustments set up the timing of the SW 30-Hz pulse to switch the video head outputs correctly. The SW 30-Hz signal leaves IC601 at pin 15 and is applied to the head-switching circuits (Sec. 5-9, and Chapter 6) as well as the artificial V-pulse generator (Sec. 5-7).

Note that RT603 advances the odd-field switching point 6.5H from the "front porch" of the vertical sync signal, while RT602 does the same for even-field switching point.

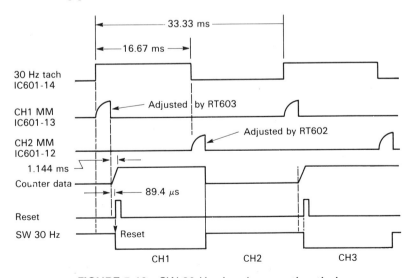

FIGURE 5-18 SW 30-Hz signal generation timing

Sec. 5-4   Cylinder Control   165

5-4.12 Cylinder Phase-Loop Troubleshooting

If there are noise bars floating through the picture on both the EVF and TV (the most common symptom for trouble in the cylinder phase-loop circuits), and the condition cannot be corrected by adjustment of the tracking controls (Sec. 5-5), play a known good tape, or a test tape (Chapter 3). Then try correcting the condition by adjustment of the cylinder phase-control (phase shifters) RT602/RT603 as described in Chapter 8.

As a next step, you can try correcting the problem by troubleshooting the speed loop as described in Sec. 5-4.5. However, if the horizontal sync is good on both the EVF and TV, the cylinder speed loop is probably good. Proceed as follows.

Place the VCR in record, and check for 1748-Hz squarewaves at pin 29 of IC601 (Fig. 5-15). If missing, with 60-Hz vertical sync pulse at pin 17, suspect IC601.

Check for 30-Hz control-track record signals at pin 18 of IC601. If missing, with 60-Hz sync pulses at pin 17, suspect IC601. It is also possible that the record signals are being inhibited by a high at pin 11 of IC601. The inhibit signal is produced by the system-control of microprocessor IC901 (Chapter 7) when the tape transport is in load/unload or when the safety tab is removed from a cassette (and the safety-tab switch moves to the no-tab position).

If the 60-Hz sync pulses are missing at pin 17 of IC601, suspect the LPF or IC203 in the video signal-processing circuits (Chapter 6).

If the 3.58-MHz signals are missing at pin 25 of IC601, suspect C608 or IC204 in the video signal-processing circuits (Chapter 6).

If the 30-Hz signals are missing at pin 14 of IC601, suspect IC610 (Fig. 5-16). To check operation of the 30-Hz PG pulse generator in IC610, check for 30-Hz signals at pin 3, 360-Hz pulses at pin 17, and 45-Hz pulses at pin 12. If the 45-Hz and 360-Hz input pulses are missing, trace back to the source. If the inputs are present, but there is no output at pin 3, suspect IC610. If there is an output at pin 3 of IC610, but the frequency is not correct, suspect the time-constant network C647/R687 at pin 14.

Place the VCR in play, and check for 30-Hz pulses at pin 15 of IC601. If missing, suspect defective PG shifter adjustment networks at pins 12 and 13 of IC601 (or a missing 30-Hz pulse at pin 14). If the networks appear to be good, and there is a 30-Hz signal at pin 14, but no output at pin 15, suspect IC601.

Place the VCR in stop, and check for a 50% duty cycle of the signal at pin 29 of IC601. Also check for a d-c signal of about 2.8 V at the junction of R636 and the LPF.

If all the signals described thus far appear to be normal, you can make a *rough check of cylinder phase-loop operation as follows.* In either record or play (or both), monitor the pulses at pin 29 of IC601. Check the duty cycle. Then (very carefully) apply slight pressure with your finger to the top of the rotating cylinder. This should cause a change in the duty cycle of the PWM output at pin 29. If the duty cycle varies when pressure is applied, the cylinder phase loop is responding correctly. If there is no variation in duty cycle with pressure applied, it is possible that IC601 is defective or is receiving a signal that locks the PWM to produce a fixed duty cycle. For example, pin 28 may be driven high by a signal from search, cylinder on, or the free-run arm test point.

5-5 CAPSTAN CONTROL

Figure 5-19 shows the capstan servo circuits in block form. IC603 controls speed, while IC601 controls phase of the capstan motor. Output signal from IC601 and IC603 are added and then applied to the capstan motor through drive circuit IC604.

Speed-control IC603 samples the 720-Hz capstan FG pulse and produces the corresponding speed-control output at pin 15.

Phase-control IC601 compares the 3.58-MHz signal from the video circuits with the 30-Hz CTL pulse from the control head (during play), or with the capstan FG pulse from the capstan motor (during record), to produce the

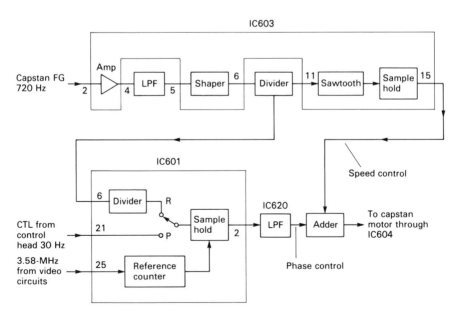

FIGURE 5-19 Capstan servo circuits

Sec. 5-5   Capstan Control   167

corresponding phase-control output at pin 2. This output is applied through an LPF in IC620 to the adder circuit.

5-5.1 Capstan Speed Control

Figure 5-20 shows typical capstan speed-control circuits.

The capstan speed-control servo system is very similar to that of the cylinder speed-control system (Sec. 5-4) and to that of many VHS VCRs. The capstan speed-control circuit is primarily contained in IC603 (the same as the cylinder speed-control circuits are in IC602). Both ICs are of the analog sample-and-hold type.

Since most camcorders are single-speed machines (with capstan speed fixed at 30 rps, except during search fast forward and rewind, as discussed in Chapter 7), there is no need for an autospeed select system as there is for many modern VCRs. In any event, only one reference or sample signal is needed for speed control. The capstan FG signal (which is 720 Hz during

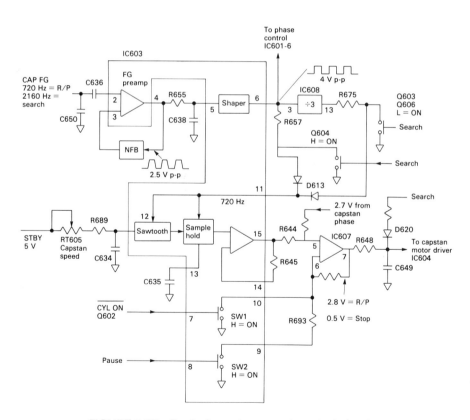

FIGURE 5-20 Typical capstan speed-control circuits

normal record/play and 2160 Hz during search) is used as the reference or sample.

During record and normal play, the capstan FG signal is applied to the FG preamp, which develops an output signal at pin 4 of IC603. The output signal is routed through a low-pass filter C638/R655 to a shaper in IC603. The shaper output of 720 Hz is applied through R657 and D613 to pin 11 of IC603.

During search operations, the capstan motor rotates at three times normal speed (90 rps instead of 30 rps), and the capstan FG pulses are at 2160 Hz. These 2160-Hz signals at pin 6 of IC603 are routed through divider IC608 where the signals are divided down to 720 Hz and applied to pin 11 of IC603 through R675 and D613.

Note that when search is selected, the system-control microprocessor IC901 makes the search line high. This turns Q604 on and shorts R657 to ground, preventing 2160-Hz signals from reaching pin 11 of IC603. When normal record/play is selected, the search line is low. This turns Q603/Q606 on and shorts R675 to ground, preventing the output of IC608 from reaching pin 11 of IC603. Since Q604 is off with normal record/play, the output at pin 6 of IC603 is passed through R657 to pin 11.

The 720-Hz signal at pin 11 of IC603 is applied to the sawtooth generator and sample-and-hold circuits in IC603. The sawtooth generator time-constant network connected at pin 12 of IC603 consists of C634 and the combination of R689 and the capstan speed-adjust control RT605. The sample-and-hold circuit filter C635 is connected at pin 13. The resultant d-c sample or control voltage is amplified and exits IC603 at pin 15.

The capstan speed-control is applied through R645 to the input at pin 5 of IC607, along with the phase-control voltage through R644. The combined phase- and speed-correction voltages are amplified by IC607, and a resultant d-c voltage at pin 7 is applied through a low-pass filter C649/R648 to the capstan motor driver IC604. An additional voltage is applied to the capstan motor drive through D620 and a resistor network during search. The additional voltage increases capstan motor speed during search (from 30 rps to 90 rps).

Also associated with the voltage applied to the input at pin 5 of IC607 are two switches within IC603. These two switches are controlled by the pause line and by the cylinder-on line at pins 8 and 7, respectively, of IC603.

During the period when the cylinder-on line is activated (stop, fast forward, rewind, etc.), the switch within IC603 is turned on, pulling pin 10 to ground. This grounds the input to pin 5 of IC607, turning off the drive voltage to capstan motor driver IC604. During the STOP mode, the voltage output at pin 7 of IC607 is about 0.5 V (compared to about 2.8 V during record/play).

During the PAUSE mode, the switch in IC603 pulls pin 9 of IC603 low, grounding R693. This shifts the d-c voltage level at the input of IC607 in the PAUSE mode.

5-5.2 Capstan Speed-Loop Troubleshooting

The most common symptoms of problems in the capstan speed-loop are excessive wow and flutter, picture noise, or picture instability (out of sync), particularly vertical instability. Play a known good tape or a test tape (Chapter 3). Then try correcting the condition by adjustment of the capstan speed control RT605 (Fig. 5-20).

As discussed in Chapter 8, the first step in adjusting RT605 is to disable the phase-control circuits (so that the phase loop will not try to lock up and correct a malfunction in the speed loop). You then adjust RT605 until a sampling pulse is stationary (or moves slowly) on the PG pulse. This should restore picture stability.

As a next step, you can try correcting the problem by troubleshooting the phase loop as described in Sec. 5-5.4. However, if the capstan motor speed can not be locked by adjustment of RT605, suspect IC603 and the associated circuits. Proceed as follows.

First, check for the capstan FG signal at pin 6 of IC603. If missing, suspect a defective FG generator in the capstan motor, IC603, or the low-pass filter (C638/C655).

Next, check for 720-Hz signals at pin 11 of IC603. If missing, suspect a defective R657, Q604, or D613.

Next, check for a d-c signal of about 2.6 V at pin 15 of IC603. If missing, suspect a defect with the sawtooth generator network connected at pin 12 of IC603, or the sample-and-hold capacitor C635 at pin 13, or a circuit within IC603.

Next, check for a d-c signal of about 2.8 V at pin 7 of IC607. If absent or abnormal, suspect IC607 or a missing reference voltage at pin 6 of IC607.

If the normal voltage is obtained at pin 7 of IC607, the capstan speed loop is operating correctly. Under these conditions, if the capstan motor is not rotating at the correct speed, the problem is likely in IC604 or the capstan motor itself.

Note that when the VCR is in stop, the voltage at pin 7 of IC607 is about 0.5 V rather than the 2.8 V during record/play.

5-5.3 Capstan Phase Control

Figure 5-21 shows typical capstan phase-control circuits.

The capstan phase-control servo system is very similar to that of the cylinder phase-control system (Sec. 5-4) and to that of many VHS VCRs. The capstan phase-control circuit is primarily contained in IC601 (the same

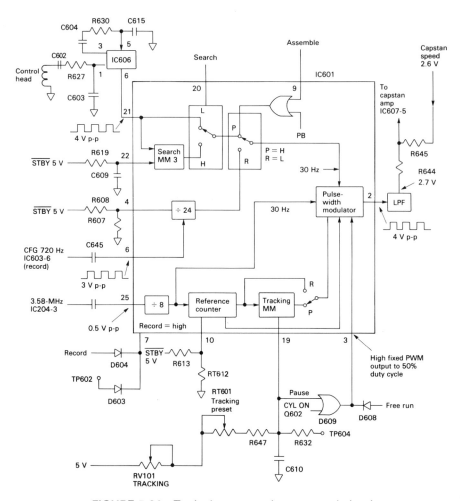

FIGURE 5-21 Typical capstan phase-control circuits

as the cylinder phase-control circuits are in IC601). Again, pulse-width modulation is used. As with the cylinder phase-loop system, a reference signal is taken from the 3.58-MHz chroma oscillator signal at pin 25 of IC601. This signal is divided down to become the reference signal for the pulse-width modulator.

The divider output is passed into a tracking multivibrator to provide a variable delay. The delay time of the reference signal is varied by an adjustable network present at pin 19 of IC601. This network consists of C610, resistors R647 and RT601 (an internal tracking preset control), and thumbwheel control RV101 (an external operating control).

The resistors in the timing network at pin 19 of IC601 return to a 5-V source. The impedance from pin 19 to the 5-V source determines the time

Sec. 5-5 Capstan Control

constant for the delay of the tracking multivibrator. This allows for shifting the timing of the reference signal to compensate for variations on various videotapes. (This is the familiar tracking function found on most VCRs.)

In addition to the divided-down reference signal (from pin 25 of IC601), the phase-control circuits use an additional comparison signal to provide control of the capstan motor. This comparison signal is taken from one of two sources, depending on the operating mode (playback or record).

During playback, the comparison signal is the 30-Hz control track pulse (CTL) from IC606, applied at pin 21 of IC601. During normal playback, the CTL pulses are applied directly to the pulse-width modulator. During search, the CTL pulses are divided by 3 (since the tape is moving at three times normal speed). This is controlled by the search signal applied to pin 20 of IC601. In either normal or search playback, the pulse-width modulator receives 30-Hz comparison signals.

The pulse-width modulator compares the phase of the 30-Hz comparison signal to the divided-down 30-Hz reference signal, generating a variable pulse-width squarewave output signal at pin 2 of IC601. The squarewaves are passed through a low-pass filter circuit which extracts only the d-c component of the variable squarewaves. The resulting d-c control voltage is applied through R644 to the input of IC607, along with the capstan speed-control voltage through R645 (as shown in Fig. 5-20).

During record, the comparison signal is the 720-Hz capstan FG signal (CFG), which is derived from the shaper at pin 6 of IC603 (Fig. 5-20). As shown in Fig. 5-21, the CFG signal is divided down by 24 to produce the required 30-Hz comparison signal during record.

The pulse-width modulator operates the same way for both playback and record (producing squarewaves with variable pulse width or duty cycle), with one minor exception. During record, the variable tracking multivibrator is bypassed. This fixes the delay and defeats the front-panel tracking control RV101.

As in the case with the cylinder phase loop, the capstan phase loop has a control input at pin 3 of IC601 to force the duty-cycle output to 50% during the pause and cylinder-on (stop, rewind, fast forward, etc.) modes. When pin 3 is forced high, either by pause/cylinder-on signals (through D609) or by connecting the free-run test point to B+ (through D608), the duty cycle of the squarewave pulses at pin 2 is fixed at 50%. As discussed in Chapter 8, the free-run test point is used during adjustment of the capstan speed-loop circuits.

5-5.4 Capstan Phase-Loop Troubleshooting

The most common symptom of problems in the capstan phase loop is a noise bar (or bars) floating through the picture (on both EVF and TV), but with the picture properly synchronized. (If the capstan speed loop is malfunctioning, you will get excessive wow and flutter, along with picture instability, out of vertical sync, etc.) Play a known-good tape or a test tape (Chapter 3). Then try correcting the condition by adjustment of the tracking controls RT601/RV101 as described in Chapter 8.

As a next step, you can try correcting the problem by troubleshooting the speed-loop as described in Sec. 5-5.2. However, if the vertical sync is good on both the EVF and TV, and there is no wow and flutter, the capstan speed loop is probably good. Proceed as follows.

Place the VCR in playback, and check for variable duty-cycle squarewaves at pin 2 of IC601. If missing, suspect IC601.

Before you condemn IC601, check for 30-Hz CTL pulses at pin 21. If missing, suspect IC606. Also check for 3.58-MHz signals at pin 25. If missing, trace back to the source (IC204, Chapter 6).

Place the VCR in record, monitor the PWM signal at pin 2 of IC601, and apply *light* pressure with your finger to the rotating capstan motor. A variation in duty cycle of the squarewaves should be seen when the pressure is applied.

If there is no change in duty cycle with pressure, suspect IC601. However, first check that pin 3 of IC601 is low (so that the PWM can operate in the variable mode). If pin 3 is high, check for a defect in the pause or cylinder-on signal sources (through D609), or for a high at the free-run test point (through D608). Also check that pin 7 of IC601 is high. If not, suspect D604 or the record-control line. Finally, check that 720-Hz CFG pulses are available at pin 6 of IC601 during record. If not, track back to the source (IC603).

5-6 CTL PULSE RECORDING/PLAYBACK PATH

Figure 5-22 shows the CTL pulse path during both record and playback. As discussed in Sec. 5-5, the CTL pulse is used by the capstan phase-control circuit during playback.

During record, the $\frac{1}{2}$-V sync signal (Sec. 5-4.9) is recorded on the tape by the control head as the CTL pulse. The CTL pulse is applied through record/inhibit circuits within IC601 and is output to the control head at pin 18 of IC601. The record/inhibit circuits are controlled by load and pause signals (through D601 to pin 11) and the record signal (to pin 7). If either load or pause go high, the record/inhibit circuits are off (open) and the CTL pulse does not reach the control head. If the record input is high, the record/inhibit circuits are on (closed), and the CTL pulse is applied to the control head.

During playback, the CTL pulses (recorded on tape and detected by the control head) are applied to IC606 through LPF C603/R627. The CTL pulses are amplified by IC606 and are applied through LPF C615/R630 to a shaper within IC606. The shaped pulses fall at the negative swing and rise

Sec. 5-7   Artificial V-Pulse Generator   173

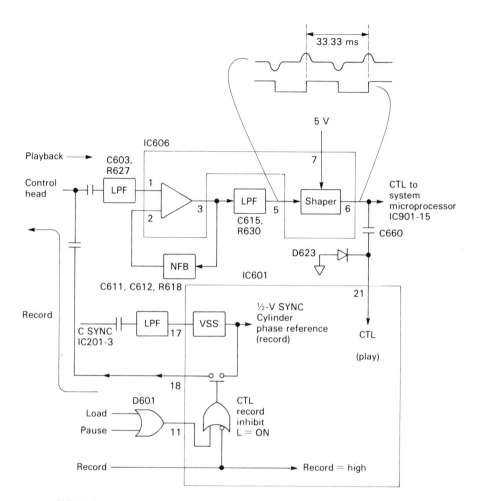

FIGURE 5-22   CTL pulse path during both record and playback

at the positive swing, of the 30-Hz CTL pulse, as shown by the timing diagram of Fig. 5-22.

The CTL pulse is applied to the capstan phase-control circuits in IC601 as the comparison signal during playback. The CTL pulse is also used by the system-control microprocessor IC901 (Chapter 7). IC901 counts CTL pulses on the trailing edge and produces corresponding ASBL (assemble) pulses to check the *phase match edit* (Sec. 5-8).

5-7 ARTIFICIAL V-PULSE GENERATOR

Figure 5-23 shows the artificial V-pulse generator circuits. Most of these circuits are contained within IC609. The circuits are used during "trick play"

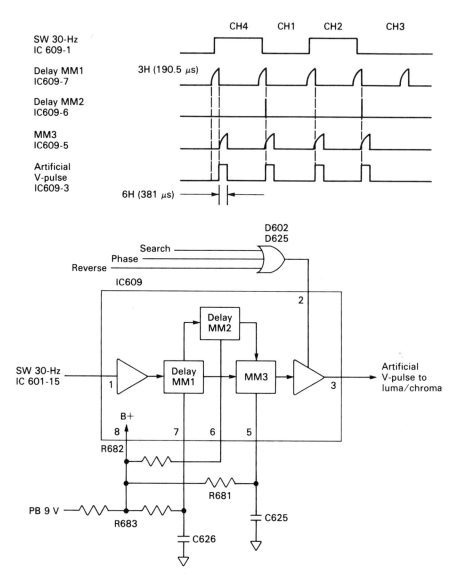

FIGURE 5-23 Artificial V-pulse generator circuits

(visual search, still, and review modes) to minimize vertical distortion or jitter on the screen (EVF and TV). Such jitter can occur when there is noise in the area of the vertical sync. IC609 produces an artificial vertical-sync pulse during trick play modes.

The generator circuits consist of three monostable multivibrators. Multivibrator 1 (MM1) determines the position of the channel 2 and channel 4 vertical sync pulse, and multivibrator 2 (MM2) determines the position of the

Sec. 5-8   Phase Matching   175

channel 1 and channel 3 vertical sync pulse. Multivibrator 3 (MM3) determines the width of the artificial V-pulse.

The artificial V-pulse is derived from the SW 30-Hz signal (Sec. 5-4.11) applied at pin 1 of IC609. The V-pulse output is at pin 3 of IC609. This output is applied to the luminance/chroma circuits (Chapter 6).

5-8   PHASE MATCHING

Figure 5-24 shows the phase-matching (or phase match edit) circuits in block form. These circuits reduce *frame joints* that can occur if the system is moved from record to pause and then back to record. During such operations, a blank or joint can appear on the tape. This blank is seen as *noise in the picture* during playback. To prevent the blank, a phase match edit operation is performed automatically by the circuits of Fig. 5-24 each time the record-to-pause-and-back-to-record sequence takes place. The edit must be performed with considerable accuracy. If not, and the blank is eliminated by simply overlapping the joints, a *rainbow effect* or *rainbow interference* can occur.

When the system-control microprocessor IC901 detects a record-pause-record sequence, IC901 outputs a high from pin 57 to interrupt video signal

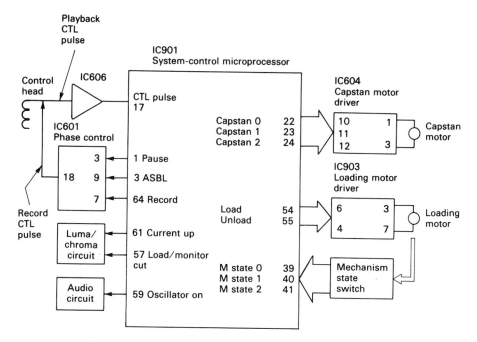

FIGURE 5-24   Phase-matching (phase-match edit) circuits

recording. (Pin 57 of IC901 also goes high during load.) Pin 59 of IC901 goes low to interrupt audio recording during the record-pause-record sequence. Simultaneously, pin 1 of IC901 goes high. This high is applied to pin 3 of servo phase-control IC601 and interrupts the CTL pulse recording. Pins 22–24 of IC901 produce signals to the capstan motor drive to reverse tape direction. (Operation of the capstan motor drive is discussed further in Chapter 7.)

As the tape moves in the reverse direction, CTL pulses are applied to pin 17 of IC901 through IC606. IC901 then counts the pulses. When the count reaches 40, IC901 outputs an unloading signal from pins 54–55 to the loading motor driver IC903. This releases the pressure roller (Secs. 5-1 and 5-2) from the capstan shaft. (Operation of the loading motor drive is discussed further in Chapter 7.) The capstan motor then receives signal from IC901 to return the capstan motor to the forward direction. However, the tape remains stationary, since the pressure roller is still disengaged.

When the system is moved from record/pause to pause, IC901 outputs a signal from pins 54–55 to reengage the pressure roller with the capstan shaft. The mechanism-state (or mode-sense) switch sends signals to pins 39–41 of IC901, indicating that the loading operation is complete and that the pressure roller is engaged with the capstan shaft. (Operation of the mechanism-state switch is discussed further in Chapter 7.)

The pause signal at pin 1 of IC901 is removed. However, because the ASBL signal (Sec. 5-6) is still present at pin 3 of IC901, recording remains interrupted. The ASBL output pulses from pin 3 of IC901 are applied to phase control IC601 and prevent recording of the CTL pulses. The ASBL pulses also match the video head and CTL pulse phases, as well as the video head and sync signal phases.

Playback CTL pulses are input to pin 17 of IC901 and are counted down to determine the amount of tape overlap. When the count reaches 34, the ASBL output from pin 3 and the load/monitor cut output from pin 57 are removed. Simultaneously, pin 59 outputs a high or oscillator-on signal to start recording, and pin 61 outputs a high or current-up signal (for approximately 1 sec). This 1-sec increase in video-signal record current reduces rainbow noise caused by the overlap of recorded signals.

VIDEO/AUDIO SIGNAL-PROCESSING CIRCUITS

This chapter describes the theory of operation for typical camcorder video/audio signal-processing circuits. All the notes described in the introduction of Chapter 4 apply here.

6-1 VIDEO/AUDIO SIGNAL-PROCESSING BASICS

The video/audio signal-processing circuits of a VHS camcorder are very similar to those of a VHS VCR, with three major exceptions:

First, the video for a camcorder is usually taken from the camera section (Chapter 4) rather than from a tuner (as is the case with a VCR). On most camcorders, it is possible to receive video from other sources (character generator, VCR, etc.). The video from the camcorder is applied to the EVF as well as to the TV. Because of these factors, special video in/out-selection circuits are required for a camcorder.

Next, the audio for a camcorder is usually taken from the microphone, although it is possible to receive audio from an external source. This requires special audio selection circuits.

Finally, because of the 4-head video recording configuration, special head-switching circuits are required for a VHS camcorder.

As a result of these circuit differences, we concentrate on video in/out-selection, audio-selection, and special head-switching circuit in this chapter.

However, we also review basic VHS luminance/chroma record/playback operation.

Keep in mind that the luma/chroma circuits for VHS camcorders are contained in three or four ICs, as is the case with most VHS VCRs. If all else fails, you can replace the few ICs, one at a time, until the problem is solved. The one major exception to this applies to the adjustment of controls in video circuits. The adjustment controls are found outside the ICs. However, when you go through the adjustment procedures recommended in the service literature, you simultaneously localize faults in the adjustment-control circuits.

Also remember that each camcorder has its own unique set of circuits which you must check out during troubleshooting (using the service literature, it is hoped). However, to give you a head start in video-circuit troubleshooting, we conclude the chapter with overall luminance and chroma troubleshooting tips that apply to virtually all camcorders.

6-2 VIDEO IN/OUT-SELECTION CIRCUITS

Figure 6-1 shows typical video in/out-selection circuits.

In this particular camcorder, the video applied to the VCR-section luma/chroma-processing circuits can be taken from one of three sources: (1) from the camera-section video output (Chapter 4), (2) from an optional external character generator attached to the camcorder via the EVF connector, and (3) from an optional external audio/video (A/V) input adapter (which also uses the EVF connector).

In our camcorder, if either of the optional input systems is used, the EVF must be unplugged from the camera. In the case of the character generator, the generator is plugged into the EVF connector, and the EVF is plugged into the interconnect receptacle on the character generator.

When the EVF is connected to the EVF connector during normal camera recording, the base of Q1515 and D1502 are both off. Standby 5 V is applied to the base of the switching FET Q1514, turning Q1514 on. With Q1514 on, the camera video output from Q1133 is applied to the luma/chroma circuit (IC203, IC204) through CN1501-13. In turn, the luma/chroma circuit applies the video signal to the EVF through IC960.

When an external video source is applied to the EVF connector using the A/V input adapter, pin 10 of the EVF connector is grounded, and D1502 turns on. With D1502 on, Q1514 is turned off. This cuts off the camera video signal and selects the external video source (through the A/V input adapter).

Connecting a character generator to the EVF connector applies a high to pin 11 of the EVF connector, turning Q1515 on. With Q1515 on, Q1514

Sec. 6-2 Video In/Out-Selection Circuits

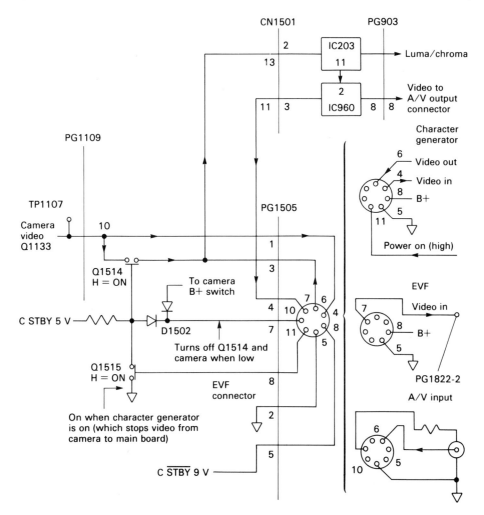

FIGURE 6-1 Typical video in/out-selection circuits

turns off, and the camera video is applied through pin 4 of the EVF connector to the character generator.

The character generator produces signals which are added to the camera video signals. The combined video signals are output through pin 6 of the EVF connector and applied to the luma/chroma circuits.

6-2.1 Video In/Out-Selection Troubleshooting

During normal camera operation, check for the presence of video from the camera at PG1109-10 or at Q1133. If missing, check camera operation as discussed in Chapter 4.

If camera video is good, check for video at CN1501-13. If missing, suspect Q1514, Q1515, or the associated components.

If there is good camera video to the luma/chroma circuits at IC203-2, check for video from the luma/chroma circuits. Start at IC960-3 and trace through to PG1822-2 on the EVF board. If the video is absent or abnormal at IC960-3, suspect the luma/chroma circuits. If the video is good at PG1822-2, but there is no picture on the EVF, suspect the EVF (Chapter 4).

If a problem occurs only when the optional character generator or optional A/V input adapter is used, check that the base of Q1514 is low and the Q1514 is off. If not, suspect D1502, Q1515, or the associated optional device. If Q1514 is off, but there is no video to IC203, suspect the associated optional device.

6-3 LUMA/CHROMA RECORD PROCESS

Figure 6-2 shows the combined luma/chroma signal-processing functions for record in simplified block form.

Video from the video in/out-selection circuits is applied to the input of both luminance record IC203 and chroma record IC204. The luminance portion of the composite video signal is converted to an FM carrier at a modulation deviation of 3.4 to 4.4 MHz. The FM luminance carrier is applied to the luminance input of IC202. The chroma portion of the composite video signal is down-converted from 3.58 MHz to 629 kHz by IC204. The 629-kHz chroma signal is applied to chroma input of IC202, summed with the FM luminance signal, and then routed to the video heads through IC201.

As discussed, VHS camcorders use four video heads with 270° tape wrap. Since each head is positioned 90° apart, more than one head is in contact with the tape at any given time. To prevent overwriting adjacent tracks during record operation, the record current signal to the video heads must be switched at precise intervals.

The signals for head switching are generated in IC610, while the actual head-switching process takes place in IC201. IC610 uses the SW 30 Hz, the 360-Hz cylinder FG, and the 45-Hz cylinder PG signals to produce the synchronized head-switching signals SW1–SW4.

6-3.1 Luma/Chroma Record Troubleshooting

As in the case of a VCR, the basic approach for troubleshooting of the luma/chroma circuits in a camcorder is to trace signals through the luma/chroma ICs. For the record process (Fig. 6-2), trace from the video in/out-selection circuits to the video heads, using the service data to locate inputs and outputs. When tracing, make certain to check any components outside the ICs. For example, IC204 has three external capacitors and an external

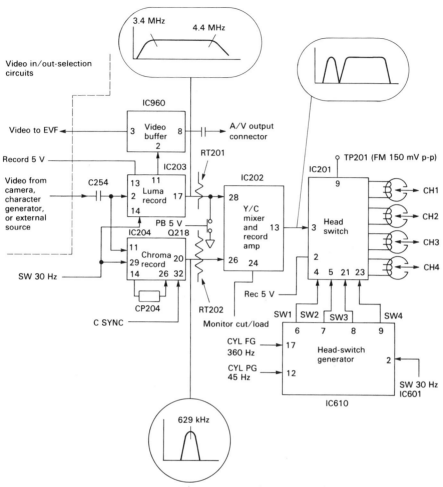

FIGURE 6-2 Combined luma/chroma signal-processing functions for record

bandpass filter, all of which are required to complete the chroma signal path within IC204.

Start by checking for a composite video signal at pin 2 of IC203 and pin 11 of IC204. If missing, suspect the video in/out-selection circuits (Sec. 6-2).

If there is good video at the input to the luma and chroma record ICs, check for video at the corresponding outputs. Trace the 629-kHz chroma signal from pin 20 of IC204 to pin 26 of IC202, through chroma record control RT202. If necessary, adjust RT202 as described in Chapter 8.

Trace the 3.4- to 4.4-MHz luminance signal from pin 17 of IC203 to pin 28 of IC202, through the luminance record control RT201. If necessary, adjust RT201 as described in Chapter 8.

If there is a good composite video signal at pin 2 of IC203, but the output at pin 17 of IC203 is absent or abnormal, suspect IC203 or the associated external components.

Also check that all the signals and voltages are available to IC203 for record. As an example, there must be SW 30 Hz signals at pin 14 to operate the FM modulator in IC203. Likewise, there must be 5 V at pin 13 of IC203 to turn on the record functions. On the other hand, pin 16 of IC203 must be low. If not, IC203 goes into PLAYBACK mode.

If there is a good output at pin 17 of IC203, but not at pin 28 of IC202, check that Q218 is off. Q218 is turned on when playback is selected. This bypasses record signals to ground during the PLAYBACK mode.

If there is a good composite video signal at pin 11 of IC204, but the output at pin 20 of IC204 is absent or abnormal, suspect IC204 or the external components.

Also check that all the signals and voltages are available to IC204 for record. As an example, there must be SW 30 Hz signals at pin 29 to operate the phase shifter within IC204. Likewise, there must be C SYNC signals at pin 32 to operate the burst gate within IC204. Pin 12 of IC204 must be low. If not, IC204 goes into the PLAYBACK mode.

If both the luma and chroma signals are present at the input of IC202, but the output at pin 13 of IC202 is absent or abnormal, suspect IC202 or the external components.

Check the status of pin 24 of IC202. This pin goes high during load and momentarily during phase-match edit operation (Sec. 5-8). If pin 24 of IC202 is high, the record output of the Y/C mixer in IC202 is shorted to ground, and the recording signal does not pass to the video heads. Likewise, the current-up signal at pin 17 of IC202 should be on only for about 1 sec, following a record/pause/record operating sequence (Sec. 5-8).

If the output from pin 13 of IC202 is good, and appears at pin 3 of IC201, but the signal is not being recorded on one or more of the video heads, suspect IC201, the video heads, or the external components.

Check for switching signals SW1-SW4 at pins 4, 5, 21, and 23 of IC201. If the switching signals are absent, suspect IC610. Also check that IC201-1 is low and IC202-2 is high. If not, IC201 may be in the playback mode. Troubleshooting for the head-switching functions of IC201 and IC610 during record is discussed further in Secs. 6-5 and 6-6.

6-4 LUMA/CHROMA PLAYBACK PROCESS

Figure 6-3 shows the combined luma/chroma signal-processing functions for playback in simplified block form.

Although the basic luma/chroma functions for playback in a camcorder are essentially the same as in a VCR, the video head-switching operation is

Sec. 6-4   Luma/Chroma Playback Process   183

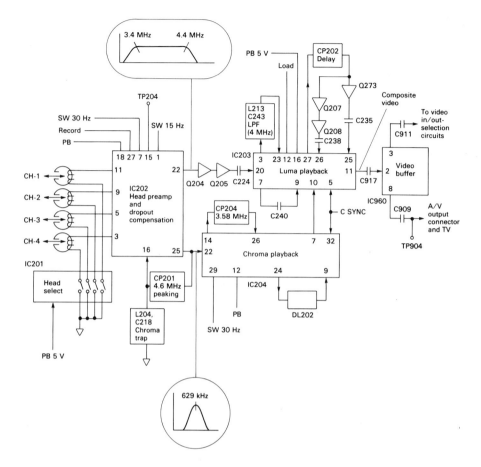

FIGURE 6-3   Combined luma/chroma signal-processing functions in playback

unique to a camcorder. The signals from four video heads must be selected in the correct sequence, and at exactly the correct time, to combine the signals properly.

During playback, IC201 grounds the return end (low end) of the corresponding video head during the period when the head is picking up (reading) the video signal on tape. Also, when one head is turned on (or active), the other three heads are turned off (shorted out). This minimizes any interference that could be developed as a result of the three inactive heads picking up signals from the tape.

The output signals from the video heads are applied to head preamplifier IC202. The SW 30-Hz and SW 15Hz signals are used by IC202 to select the correct video-head FM carrier and to produce a continuous FM carrier signal.

The FM carrier output from IC202 is filtered, and the luminance portion is passed to the input of the luminance playback circuits in IC203. The chroma playback signal, filtered out of the FM carrier signal, is applied to IC204. The 629-kHz chroma signal is up-converted to the original 3.58-MHz and coupled back into IC203. The chroma signal is mixed with the demodulated luminance FM signal in IC203 to form a composite video signal at pin 11 of IC203.

The composite video signal is applied to buffer IC960 where two video signals are developed. One signal is applied to the video in/out-selection circuits (Sec. 6-2), while the other signal is applied to the A/V output connector. (The signal at pin 3 of IC960 is applied to the EVF.)

6-4.1 Luma/Chroma Playback Troubleshooting

As with record, the troubleshooting approach for luma/chroma playback is to trace signals. For the playback process (Fig. 6-3), trace from the video heads to the outputs of IC960 (or directly to the EVF and A/V output connector).

Again, when tracing signals through the ICs, make certain to check any external components. For example, IC202 has a peaking filter, a chroma trap, a 1H delay, a phase compensator, and a buffer, all of which are required to complete the luma/chroma signal path within IC202.

Start by checking the obvious functions. For example, if playback is good on the monitor TV, but not on the EVF (or vice versa), check the outputs at pins 3 and 8 of IC960. If the signal is good at pin 3, but not at pin 8 (or vice versa), suspect IC960. (If the signal is good at pin 3, but there is no display on the EVF, suspect the EVF circuits, Chapter 4.)

If both outputs from IC960 are absent or abnormal, check for a composite signal at pin 11 of IC203 and pin 2 of IC202. If the signal is good at pin 2 of IC960, but not at pins 3 and 8, suspect IC960.

If there is no composite video at pin 11 of IC203, check for proper inputs at pins 20 and 10 of IC203. If missing, suspect IC202 and/or IC204. If present, suspect IC203.

Before you pull IC203 (or any other IC), check that all the signals and voltages are available to IC203 for playback. As an example, the playback signal at pin 16 must be high to turn on the playback function. Likewise, there must be a C SYNC signal at pin 32 of IC204 before IC204 will provide a proper signal from pin 7 of IC204 to pin 10 of IC203. Also, during load or phase-match edit operation, pin 12 of IC203 goes high to squelch the video signal.

Before you pull IC204, check that all the signals and voltages are available to IC204 for playback. As an example, IC204-12 must be high for playback, and there must be SW 30-Hz signals at IC204-29 as well as C SYNC signals at IC204-32.

It is often difficult to check the outputs of individual video heads on many camcorders (as it is on many VCRs). That is why test point TP204 is provided. TP204 monitors the continuous FM signal, after some amplification and reconstruction by IC204. However, it should be possible to monitor the 629-kHz chroma output at pin 25, and the luminance output at pin 22, of IC202.

If you can measure the outputs of the video heads (say, at pins 3, 5, 9, and 11 of IC202), the important point to remember is that the output from all heads should be substantially the same. If not, the head with the abnormal output or the corresponding switch in IC201 are suspect. We discuss troubleshooting for the video head-switching circuits during playback further in Sec. 6-7.

6-5 VIDEO HEADSWITCHING SIGNAL GENERATION

Figure 6-4 shows the signal-generator circuits for the video head-switching functions in simplified form together with the timing diagrams. Note that most of the head-switching signals are generated by circuits within IC610.

During both record and playback, the four video heads are switched into and out of the circuit during the period the heads contact the surface of the tape. This is done by the timing system in IC610. The timing system is referenced to the cylinder motor 45-Hz PG signal (one pulse per revolution) and the cylinder 360-Hz FG signal (Chapter 5).

During record, the 45-Hz PG signals and the 360-Hz FG signals are input to pins 12 and 17 of IC610, respectively. These signals are used to develop the 30-Hz PG output signal at pin 3 of IC610. The 30-Hz PG signal is then passed to IC601-14 (Chapter 5), where the signal is processed by the PG shifters to develop the SW 30-Hz signal at pin 15 of IC601. The SW 30-Hz signals are coupled back to IC610 at pin 2 and are divided down by a factor of 2. The divided-down 30-Hz signal is referred to as the SW 15-Hz signal and appears at pin 19 of IC610 for distribution to IC202 (Sec. 6-4).

As shown in the record timing diagrams of Fig. 6-4, the 30-Hz PG and delayed SW 30-Hz signals are used by IC610 to develop the four head-select signals SW1–SW4. Notice that the ending and starting of adjacent head-select signals overlap (by about 4.16 ms). This compensates for any variation in the timing of the cylinder rotation so that the video tracks recorded on tape always record longer than what is needed.

During playback, the SW 15-Hz and SW 30-Hz signals are used by IC610 to determine the head-select signals for each of the four video heads. As shown on the playback timing diagrams of Fig. 6-4, when one video head is switched off, the other video head is switched on (without overlap).

186 Video/Audio Signal-Processing Circuits Chap. 6

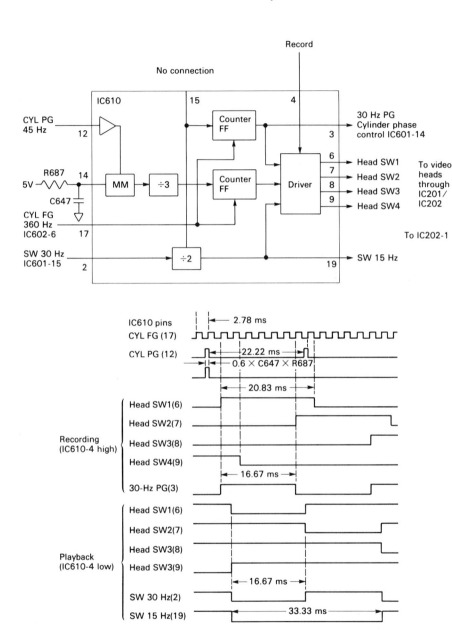

FIGURE 6-4 Signal-generation circuits for video head switching

6-5.1 Video Head-switching Circuit Troubleshooting

To check that IC610 is producing the correct signals, select the record mode, and check inputs and outputs at corresponding pins of IC610.

First check that pin 4 of IC610 is high during record and low during playback. If not, suspect the system-control microprocessor IC901 (Chapter 7). For example, if you have a problem with video head-switching signals during playback, but not record, make certain that pin 4 is low before you pull IC610. Likewise, if the problem occurs during record, but not playback, make certain that pin 4 is high before you check the outputs from IC610.

Check for the 45-Hz PG signal at pin 12 of IC610. If missing, suspect a defective PG pickup sensor within the cylinder motor.

Check for the 360-Hz PG signal at pin 17 of IC610. If missing, suspect a defective IC602 (Chapter 5).

If both the 45-Hz PG and 360-Hz FG signals are present, check for the 30-Hz PG signal at pin 3 of IC610. If missing, suspect IC610 or possibly C647/R687.

Check for the SW 30-Hz signal at pin 2 of IC610. If missing, suspect IC601 (Chapter 5). If present, check for the SW 15-Hz signals at pin 19. If missing, suspect IC610.

Finally, assuming that all other signals are present, check for the four head-select signals SW1–SW4 at pins 6, 7, 8, and 9 of IC610. If any head-select signal is absent or abnormal, suspect IC610.

6-6 VIDEO HEAD-SELECT OPERATION

Figure 6-5 shows the video head-select circuits in simplified form. Note that only one channel is shown and that most of the circuits are found within IC201. Three more identical circuits are included in IC201 to select the other three heads.

The purpose of IC201 is to select the proper video head during both record and playback. In playback, IC201 grounds the low side of the selected video head, allowing the high side to pass the signal to the playback preamp IC202. IC201 also *shorts across* the nonselected heads during playback, preventing signals from these heads from being picked up on the tape. In record, the selected head has the playback side grounded through a 10-ohm resistor (used to measure record current). The other side of the head is connected to the FM record signal. The nonselected heads are *shorted to ground* in record, preventing signals from these heads from being recorded on tape.

To help understand the record/playback video head-select operation, pay particular attention to the mode chart shown on Fig. 6-5 (at pin 4 of IC201).

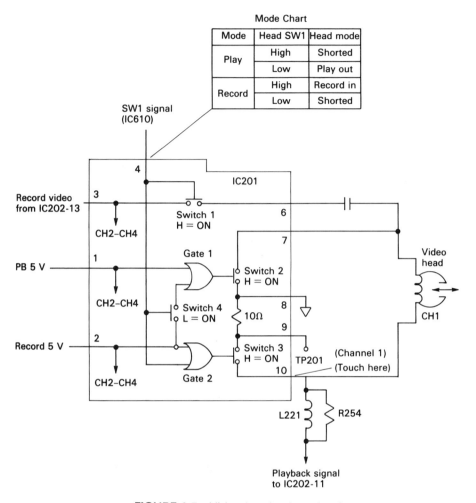

FIGURE 6-5 Video head-select circuits

6-6.1 Video Head Select During Playback

During playback, pin 2 of IC201 goes low, and pin 1 goes high. This high is applied to switch 2 within IC201 through gate 1. Switch 2 is turned on, connecting one end of the channel 1 video head to ground through pins 7 and 8 of IC201.

When pin 4 goes low during playback (when the SW1 signal from IC610 is present), switch 3 turns off (since both inputs to gate 2 are low). The output signal from the video head at IC210-10 is applied to the corresponding playback amplifier within IC202 (as discussed in Sec. 6-7).

When pin 4 goes high during playback (when the SW1 signal from IC610

Sec. 6-6   Video Head-Select Operation   189

is not present), switch 3 turns on, shorting the other end of the channel 1 video head to ground through pins 8–10 and the 10-ohm resistor. This prevents output from the video head, since the opposite side of the head also remains shorted to ground through pins 7 and 8 and switch 2.

Any signal available at pin 3 of IC201 is prevented from passing to the head because of the short at pins 7 and 8 (even though switch 1 is on). As a result, signals are played back by the channel 1 video head only when the SW1 signal from IC610 is present.

6-6.2   Video Head Select During Record

During record, pin 1 of IC201 is low, while pin 2 goes high. The high at pin 2 is applied to switch 3 through gate 2, turning switch 3 on. This connects one end of the video head to ground through pins 8–10 and the 10-ohm resistor.

When pin 4 of IC201 goes high during record (when the SW1 signal from IC610 is available), switch 1 turns on. The record signal at pin 3 of IC201 is applied to the other end of the video head through switch 1, pin 6 of IC201, and the capacitor.

When pin 4 of IC201 goes low during record (when the SW1 signal from IC610 is not available), switch 1 turns off, removing the record signal at pin 3 from the video head. The low at pin 4 is also applied to switch 4, turning switch 4 on.

The high at pin 2 is applied to switch 2 through switch 4 and gate 1. This turns switch 2 on, connecting the high end of the video head to ground through pins 7 and 8. The other end of the video head remains connected to ground through switch 3, pins 8–10, and the 10-ohm resistor.

As a result, no signals are recorded by the channel 1 video head during the time when SW1 is not present (when SW2, SW3, or SW4 are on). Signals are recorded on channel 1 only when SW1 is on.

6-6.3   Video Head-Select Troubleshooting

The most practical means of determining if the head-select functions are normal (during record) is to monitor the record current through the 10-ohm resistors for each video head. Actually, you monitor the voltage across the resistors and calculate the current.

Typically, the record current should be about 15 mA when a 1-V p-p composite video signal is applied at the A/V input connector. This produces a 150-mV p-p signal across the 10-ohm resistor. As discussed in Chapter 8, you can apply a 1-V NTSC video signal at the A/V input connector (using an NTSC colorbar generator) and then adjust R201/R202 (Fig. 6-2) for the correct current.

From a troubleshooting standpoint, the important point to remember is that the record current (or voltage reading across the resistor) should be *substantially the same for all four video heads.* If not, suspect IC201, IC610, or the video heads.

If the record current is the same for all four video heads, but too high or too low, the problem is probably the result of improper adjustment. If there is no record current on any video head, with all signals present at IC201, the problem is probably in the other luma/chroma circuits (IC202, IC203, IC204, video in/out-select, etc.).

The most practical means of checking the head-select functions (during playback) is to monitor the playback signals at TP204 (Fig. 6-3) with an oscilloscope. Then touch the corresponding pin on IC201 (Fig. 6-5) for each video head with a metallic screwdriver. For the channel 1 video head, touch pin 10 of IC201, as shown in Fig. 6-5.

As the pin for each head is touched, there should be some noise on the oscilloscope display, and the noise level should be about the same for each head. If the noise is substantially the same for all heads, the playback amplifiers in IC202 are probably good. Under these conditions, any playback problems are probably the result of defective heads or head-select functions.

Keep the following in mind. If the heads and head-select functions are good during record (proper record current for all heads), the heads are probably good. We discuss video head playback troubleshooting further in Sec. 6-7.

6-7 VIDEO HEAD PLAYBACK-SELECT OPERATION

Figure 6-6 shows the video head playback-select circuits in simplified form. Note that the preamplification and switching for all four channels in IC202 are shown but that switching for only one channel in IC201 is given.

The selected video head signal is coupled into the inputs of IC202 at pins 11 (channel 1), 9 (channel 2), 5 (channel 3), and 3 (channel 4). Within IC202, the video signal for each head is amplified by a low-noise amplifier. The amplified video head FM signal is passed to a set of three signal switches, toggled by the head-switching signals at pins 1 and 7 of IC202. These signals are the SW 30-Hz from IC601-15 and SW 15-Hz from IC610-19 and can be monitored at TP205 and TP202, respectively.

The FM carrier from each head is selected as an output signal that exits IC202 at pin 25. The FM carrier signal is routed to the input of the chroma process circuits in IC204 and to a filter circuit. The luminance portion of the FM signal is separated by the filter and returned to IC202 through pin 16.

After amplification, the luminance portion of the FM signal is applied

Sec. 6-7   Video Head Playback-Select Operation   191

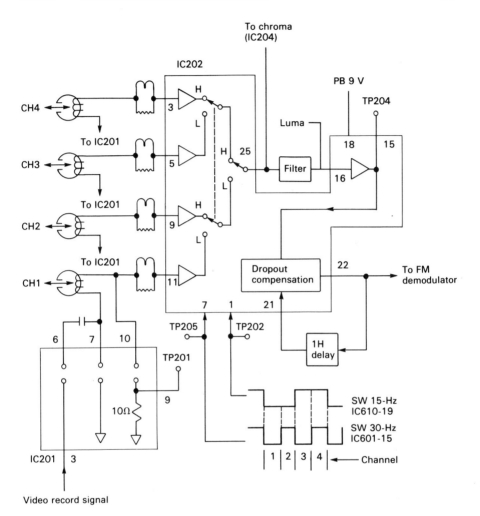

FIGURE 6-6 Video head playback-select circuits

to the FM demodulator through pin 22 of IC202. The FM carrier at pin 15 of IC204 can be monitored at TP204.

The FM carrier is also applied to a dropout-compensation circuit or DOC within IC202 and to an external 1H-delayed network (external to IC202). These internal and external components replace one or two horizontal lines of FM carrier in event of a signal dropout. The dropout circuits in VHS camcorders are identical to those for VHS VCRs.

As shown by the timing charts on Fig. 6-6, the FM video signal appearing at TP204 is taken from the channel 1 head, when both SW 30-Hz and SW 15-Hz signals are low. When SW 30-Hz goes high, the video signal at TP204 is taken from the channel 2 video head, and so on. This process is repeated for all four video heads in turn.

Keep in mind that IC201 (Sec. 6-5) selects the RECORD or PLAYBACK mode for each video head, while IC202 selects the *playback signal* from the same video head selected by IC201.

6-7.1 Video Head Playback/Select Troubleshooting

Once you are certain that all video heads are operating properly during record (with proper record current as discussed in Sec. 6-6.3), the first step in checking the playback/select functions is to monitor the FM signal at TP204 (about 100 mV p-p) with an oscilloscope.

If there is no signal at TP204 during playback, suspect IC202, or the signals at pins 1 and 7 of IC202.

If there is an FM carrier at TP204, but one of the head outputs is missing (as shown by a choppy FM carrier display), suspect IC202, IC201, or the corresponding video head.

If the FM carrier is good at TP204, check for a signal at pin 22 (about 60 mV p-p). If absent or abnormal, suspect IC202.

If there is a good signal at pin 22 of IC201, but video playback is abnormal, the video heads and switching circuits are probably good, and the problem is in the video signal-processing circuits that follow IC202. Refer to Secs. 6-9 and 6-10.

6-8 AUDIO SIGNAL-PROCESSING CIRCUITS

Figure 6-7 shows typical audio signal-processing circuits in simplified block form. Note that most of the audio amplifier circuits are contained within IC401, while the audio-switching circuits are contained in IC402.

6-8.1 Audio Input/Output Circuits

Figure 6-8 shows typical audio input/output circuits.

The audio signal is input to the audio circuit from the built-in microphone, an external microphone, or the A/V input adapter. When the A/V input adapter is used, a circuit in the adapter reduces the audio input signal level to an amplitide that equals the internal or external microphone.

The audio signal is applied to the amplifier circuits within IC401 at pin 23 through the EVF connector. The signal is amplified by circuits within IC401 and applied to the accessory earphone through JK401 as well as to external circuits (monitor, RF modulator, etc.) through the A/V output connector.

Sec. 6-8   Audio Signal-Processing Circuits

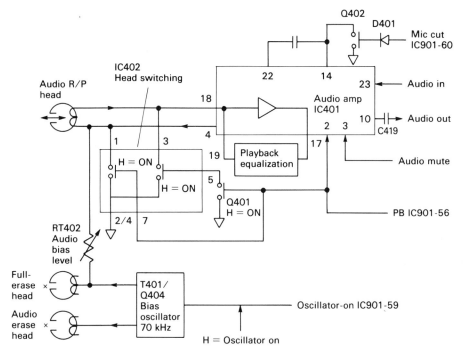

FIGURE 6-7   Typical audio signal-processing circuits

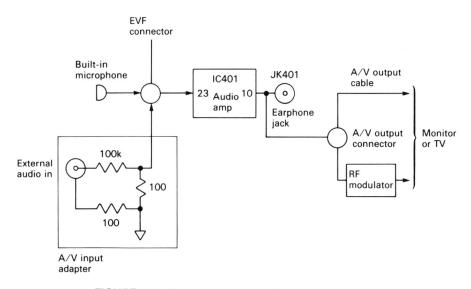

FIGURE 6-8   Typical audio input/output circuits

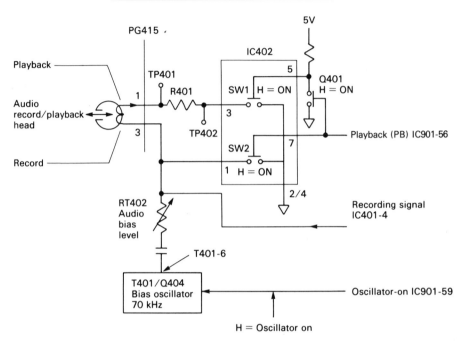

Mode	PB	Q401	IC402 SW1	IC402 SW2	Ground terminal
Record	Low	Off	On	Off	PG415-1
Playback	High	On	Off	On	PG415-3

FIGURE 6-9 Typical audio head-switching circuits

6-8.2 Audio Head-Switching Circuits

Figure 6-9 shows typical audio head-switching circuits.

The record/playback audio head is switched by the playback signal from system-control microprocessor IC901. When the playback signal is high (playback), Q401 is turned on and pin 5 of IC402 is pulled low. This turns SW1 off. The high playback signal is also applied to pin 7 of IC402, turning SW2 on. This grounds the record terminal of the audio head and allows the playback signal to be picked up by the head. The playback output from the audio head is applied to IC402 through R401.

During record, the playback signal is low. This turns SW1 on and SW2 off, grounding the playback terminal of the audio head through R401 and removing the ground from the record terminal. Under these conditions, the recording signal from IC401 is applied to the record terminal of the audio

Sec. 6-8   Audio Signal-Processing Circuits   195

head, along with the bias signal from Q404/T401. The bias oscillator Q404/T401 is controlled by an oscillator-on signal from pin 59 of IC901.

6-8.3 Audio Amplifier Circuits

Figure 6-10 shows typical audio amplifier circuits. Note that the internal circuits of IC401, as well as external circuits (feedback, equalization, traps, etc.), are shown in block form.

During playback, IC401 is placed in the PLAYBACK mode by a high (playback) at pin 2 and a low (audio mute off) at pin 3. These signals are taken from the system-control microprocessor IC901.

The audio signal from the playback terminal of the audio head is amplified (with external equalization) by the PB amplifier in IC401 and applied through the audio playback-gain control RT401 and a 15.750-kHz trap to the line amplifier in IC401. The output from the line amplifier (with external equalization) is taken from pin 10 of IC401.

During record, IC401 is placed in the RECORD mode by a low (record)

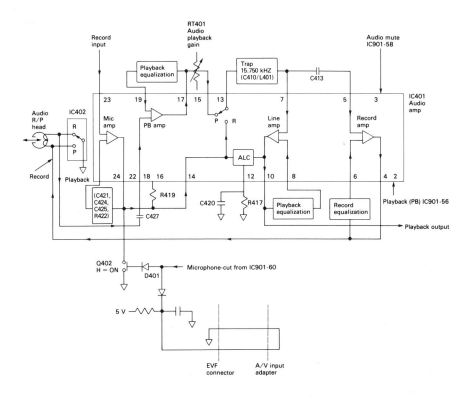

FIGURE 6-10   Typical audio amplifier circuits

at pin 2 and a low (audio mute off) at pin 3, from IC901. The audio to be recorded is applied to pin 23 of IC401. The audio is amplified (with external feedback) by the microphone amplifier in IC401 and applied through the 15.750-kHz trap to the record amplifier in IC401. The output from the record amplifier (with external equalization) is applied to the record terminal of the audio head.

The 15.750-kHz trap at pins 5, 7, and 13 of IC401 rejects any horizontal sync signals and prevents buzz during both record and playback.

The ALC (automatic level control) circuit in IC401 detects fluctuations in the input signal level and maintains the output signal level constant.

The recording amplifier equalization network uses a series resonant circuit at pin 6 of IC401 to raise the audio signal level at about 10 kHz (to provide the rated frequency response).

The playback amplifier equalization network lowers the high/medium-frequency band to about 6 dB/octave. This compensates for the increase in playback head output in the high/medium band.

IC901 applies a microphone-cut (also called an EE-cut in some camcorders) signal at pin 60 during eject operation. This signal is applied to Q402 through D401. When the signal is high (eject), Q402 turns on, shorting the amplified microphone signal at pin 22 of IC401 to ground. This prevents recording during eject.

Note that Q402 can also be turned off, even though IC901-60 is high, when the A/V input adapter is connected to the EVF connector.

Also note that the LPF connected to the microphone amplifier in IC401 (between pins 22 and 24) rejects any 70-kHz bias signal that may be picked up as the record input.

6-8.4 Audio Circuit Troubleshooting

The audio circuits for most camcorders are straightforward direct paths of input to output (from source to audio head during record and from head to output during playback). Troubleshooting for the audio system involves basic signal tracing with audio-monitor devices such as meters and oscilloscopes. For that reason we do not go into audio circuit troubleshooting in any detail. Instead, here are some tips for audio circuit troubleshooting.

Always check any adjustments in the audio system first. For example, the playback-gain adjustment RT401 must be set to provide an output of about 250 mV p-p at the A/V output connector when a 1-kHz alignment tape is played back.

Check the frequency and level of the bias signal. Usually, the bias signal is taken from the same oscillator that provides the erase signal to the erase heads (both full-erase and audio-erase). The bias oscillator frequency (typically in the 60- to 70-kHz range) is usually not critical, but the level of the bias signal (as set by RT402) is important. If the bias is too low, high frequencies are increased, resulting in distortion. If the bias is too high, high frequencies are attenuated.

It is usually difficult to measure the direct output of an audio head. However, the head output is amplified in IC401 (usually more than once). For example, the output signal during playback should be about 250 mV p-p at pin 10 of IC401.

Always check all switching functions in the audio circuit. For example, during record, the playback terminal of the audio head (Fig. 6-9) must be grounded (through SW1), while the record terminal must be connected to both the recording signal and the bias signal. During playback, the record terminal of the audio head must be grounded (through SW2), while the playback terminal must be connected to the playback input at pin 18 of IC401.

Keep in mind that the audio head must be cleaned, as is the case with the video, control track, and erase heads. Refer to Chapter 3 for cleaning procedures.

Make certain that the correct mode-control signals are applied. For example, the playback line at pin 2 of IC401 must be low for record and high for playback. The audio-mute line at pin 3 of IC401 must be low for both playback and record and should be high only during special conditions (search, load, eject, pause, rewind, stop, etc.). The same is true of the microphone-cut signal from IC901-60 applied to Q402 (Fig. 6-10).

Finally, before you decide that an audio circuit IC (such as IC401 and IC402) is defective and must be pulled, always check all connections to the IC, especially the B+ or other power connections.

6-9 LUMINANCE (BLACK AND WHITE) TROUBLESHOOTING

The first step in troubleshooting the luminance portion of the video/audio signal-processing circuits is to play back a known good tape or an alignment tape. (This is not a bad idea when troubleshooting any camcorder problem!) For one thing, you will know instantly if the problem is in the VCR section or the camera section. You will also have some hint as to whether the problem is in playback or record.

For example, if the picture and/or audio are not good during playback of a known-good tape, the problem is likely in the playback functions of the signal-processing circuits.

Next, try recording known-good audio and video from an external source through the A/V input connector. (Make sure you switch to a blank tape, not an alignment tape, *before* recording!) Then play back the recorded tape. If playback is good with the alignment tape, but not with the recorded tape, the problem is likely in the record functions of the signal-processing circuits.

If both the playback and record functions of the audio/video signal-processing circuits appear to be good, try monitoring the camera output "live" on a known-good monitor using the A/V output connector. If the "live" output from the camera is not good, you have camera circuit problems. These camera problems must be cured separately from any problems in the signal-processing circuits.

Once you are convinced that the problem is in the video/audio signal-processing circuits, run through the electrical adjustments that apply to the luminance (or Y or picture). As described in Chapter 8, there are relatively few electrical adjustments for the signal-processing circuits. A typical set of adjustments for both the luma and chroma sections includes luma/chroma separation, luma/chroma record level, audio bias level, and audio playback gain or level. Keep the following points in mind when checking performance and making adjustments.

If playback from a known-good tape has poor resolution (picture lacks sharpness), look for problems in the noise canceler (IC203, Fig. 6-3) and for bad response in the video head preamps (IC202). When making adjustments, such as the luma/chroma separation adjustment, study the stair-step or colorbar signals for any transients at the leading edges of the white bars.

If playback has excessive snow (electrical noise), try adjustment of the tracking control, including the tracking preset. Mistracking can cause snow noise. Next try cleaning the video heads (Chapter 3) before making any extensive adjustments. (As in the case of VCRs, cleaning the video heads clears up about 50% of all noise or snow problems in camcorders.) Keep in mind that snow noise can result from mistracking caused by a mechanical problem. For example, if there is any misadjustment in the tape path, snow can result. So, if you have an excessive noise problem that cannot be corrected by tracking adjustment, head cleaning, or electrical adjustment, try mechanical adjustments, starting with the tape path.

If playback of a known-good tape produces smudges on the leading edge of the white part of a test pattern (from an alignment tape) or on a picture, the problem is probably in the preamps (IC202) or in the head/preamp circuits. The head/preamp combination is not reproducing the high end (near 5 MHz) of the video signals.

If you see a herringbone (beat pattern) in the playback of a known-good tape, look for carrier leak. There is probably some imbalance condition in the FM demodulators or limiters (probably in IC203), allowing the original carrier to pass through the demodulation process. If very excessive carrier passes through the demodulator, you may get a negative picture (blacks are white and vice versa).

If the VCR section produces the correct output level when playing back an alignment tape, but not from a tape recorded on the VCR section, you may have a problem in the record circuits. As an example, *the record current may be low*. (One common symptom of low record current is snow or excessive noise.) Most camcorders have an adjustment for record level for both luma and chroma.

To sum up luminance troubleshooting, if you play back an alignment tape, or at least a known-good tape, and follow this with head cleaning and a check of the recommended alignment procedures, you should have no difficulty in locating most black and white picture problems.

6-10 CHROMA TROUBLESHOOTING

As in the case of luminance, the color portion of the video/audio signal-processing circuits is very complex, but not necessarily difficult to troubleshoot (nor do the color circuits fail as frequently as the mechanical section of the VCR in a camcorder). Again, the first step in color-circuit troubleshooting is to play back an alignment tape, followed by a check of all adjustments pertaining to color. As in the luminance circuits, when performing the adjustment procedures, you are tracking the signal through the color circuits.

There are two points to remember when making the checks. First, most color circuits are contained within ICs, possibly the same ICs as the luminance circuits. For a typical camcorder such as shown in Fig. 6-3, most of the luminance circuits are in IC203, while the chroma circuits are in IC204. Of course, the color and luminance circuits are interrelated. If you find correct inputs and power to an IC, but an absent or abnormal output, you must replace the IC. Possible exceptions in the color circuits are the various delays, filters, and traps outside the IC (on most camcorders).

Second, in the VCR section of most camcorders, the fixed input to the color converters comes from the same source for both record and playback (from crystal-controlled oscillators in IC204). If you get good color on playback but not on record, the problem is definitely in the record circuits. However, if you get no color on playback of a known-good tape, the problem can be in either the color playback circuits or the common fixed-signal source. (As a practical matter, both record and playback circuits are probably in IC204!)

The following notes describe some typical color-circuit failure symptoms, together with some possible causes.

If the hue control of the TV must be reset when playing back a tape that has just been recorded, check the color subcarrier frequency. Use a frequency counter.

If you get a "barber pole" effect, indicating a loss of color lock, the AFC circuits within IC204 are probably at fault. If practical, check that the AFC circuit is receiving the H-sync pulses and that the VCO in IC204 is nearly on-frequency, even without the correction circuits (or you can simply replace IC204).

If you get bands of color several lines wide on saturated colors (such as alternate blue and magenta bands on the magenta bar of the colorbar signal), check the 3.58-MHz oscillator frequency (and/or replace IC204).

If you get the herringbone (beat) pattern during color playback, try turning the color control of the TV down to produce a black and white picture. If the herringbone pattern is removed on black and white, but reappears when the color control is turned back up, look for leakage in both the color and luminance circuits (IC203 and IC204). For example, there could be a carrier leak from the FM luminance section (IC203) beating with color signals, or there could be leakage of the 4.27-MHz signal (IC204) into the output video.

If you get flickering of the color during playback, look for failure of the ACC system in IC204. It is also possible that one video head is bad or that the preamps are not balanced (IC202) or are not being properly switched (IC201), but such conditions also show up as a problem in black and white operation.

If you lose color after a noticeable dropout, look for problems in the burst ID circuit (within IC204). It is possible that the phase-reversal circuits have locked up on the wrong mode after a dropout. In that case, the color signals have the wrong phase relation from line to line, and the comb filter is canceling all color signals. Check both inputs (3.58-MHz input from the reference oscillator and the video input signal) applied to the burst ID circuit. If the two signals are present, check that the burst ID pulses are applied to the switchover FF.

Again, keep in mind that all the color-circuit functions discussed here may be contained within one or two ICs (probably in IC204) and cannot be checked individually. So you must check inputs, outputs, and power sources the IC as well as any external delays, filters, and traps, and then end up replacing the IC!

7
SYSTEM CONTROL and ELECTRICAL DISTRIBUTION

This chapter describes the theory of operation for typical camcorder system-control and electrical-distribution circuits. All the material in the introduction of Chapter 4 applies here.

System control is an area where one model of camcorder can differ greatly from other models. However, operation of the system-control circuits for any camcorder is determined primarily by microprocessors. Most camcorders use one microprocessor in system control, although some camcorders use two or more. In any case, the microprocessors accept logic-control signals from the camcorder operating controls and from various tape sensors. In turn, the microprocessors send control signals to video, audio, servo, and power supply circuits, as well as drive signals to solenoids and motors.

Note that many system-control functions are closely related to mechanical operation of the camcorder (VCR tape transport and servo) as well as to many circuit functions. So it is essential that you study the related chapter when reviewing system-control operation. In most cases, there is a reference from the related chapter (Chapters 4, 5, and 6) to this chapter when system control is involved.

Note that some camcorder service literature uses the term *microprocessor*, while other literature favors the term *microcomputer*. In both cases, we are discussing a single IC that performs many specific, computerlike functions in response to various inputs. The author prefers the term microprocessor and uses the term throughout this chapter. But do not be surprised to find a camcorder described as having many microcomputers in the system-control circuits or as being under control of a microcomputer.

In addition to system control, we discuss electrical distribution in this chapter, including operation of the a-c adapter and battery charger common to most camcorders.

7-1 ELECTRICAL OVERVIEW

Figure 7-1 shows overall electrical distribution and control of a typical camcorder.

Electronic control of the camcorder is provided by a single system-control microprocessor similar in concept to that found in VCRs (and in virtually all other modern consumer electronic equipment). In this case, system-control microprocessor IC901 controls the camera section, VCR section, and power-distribution circuits.

Operating power for the typical camcorder of Fig. 7-1 can be obtained from one of three sources: the adapter/charger, the Ni-Cad battery, or an external 12-V supply. The external 12-V power can be obtained from an optional automotive power cord during operation or from an external power supply during service.

Generally, the battery or adapter/charger should not be used for routine service. *An adjustable 12-V power source is preferable* for many reasons. Obviously, the battery should not be used (except during final checkout) to prevent running the battery down during long periods of service. The adapter/charger can be used during service (and should be checked at every service), but the output is not adjustable. An adjustable 12-V source can be used for extended periods of time and permits a number of special checks.

As an example, all camcorders have some form of indicator to show

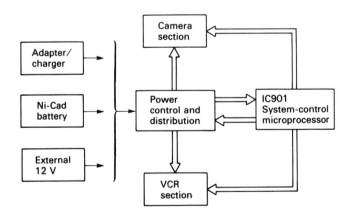

FIGURE 7-1 Overall electrical distribution and control of a typical camcorder

Sec. 7-2   AC Adapter/Charger   203

battery charge or voltage level (often called the battery overdischarge detection circuit). Some camcorders use three LEDs; others show battery condition as an on-screen display in the EVF. Either way, an adjustable power source permits you to check operation of the battery level indicator by varying the power source across the voltage range. Typically, the indicator should show full charge when the external power source is adjusted to 12.3 V (or higher) and should produce a warning (a flashing record indicator in the EVF) when the power source is at 11.62 V, and the system-control microprocessor should shut the camcorder down (pause or stop) when the source is at 10.9 V (or less).

Before we get into operation of the system-control functions, let us cover the adapter/charger circuit operations.

7-2 AC ADAPTER/CHARGER

Figure 7-2 shows typical adapter/charger circuits. This adapter/charger delivers 12-V power to the VCR/camera sections of the camcorder or is used to charge the 12-V Ni-Cad battery (but not both functions at the same time).

The adapter/charger operates on a-c voltages in the range from 100 to 240 V, at frequencies from 50 to 60 Hz. The circuits automatically compensate for variations in line voltage and frequency to maintain the rated output voltage and do not require changes in user switch settings.

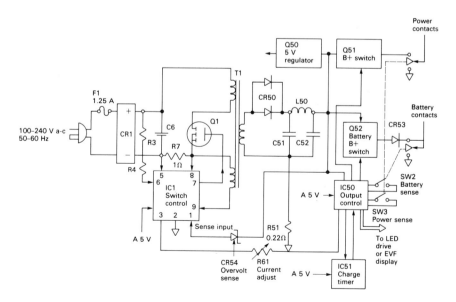

FIGURE 7-2 Typical adapter/charger circuits

7-2.1 Pulse-Width Modulation

The adapter/charger of Fig. 7-2 uses pulse-width modulation or PWM to control output voltage. Power MOSFET Q1 is switched on and off at the rate of about 100 kHz by squarewaves from IC1. This switches the current through the primary of T1 on and off at the same rate. By varying the duty cycle or pulse width of the squarewaves, the current through T1 (and thus the output voltage from the power supply) is controlled to maintain a constant output voltage.

7-2.2 Adapter/Charger Circuit

When the adapter/charger is first plugged in, power is applied to IC1 through resistors R3/R4. The start-up power causes IC1 to turn Q1 on, drawing current through the primary of T1. This induces a signal into the secondary winding of T1 that is coupled back to a sensing input of IC1. The feedback forms an oscillating system that operates at approximately 100 kHz. The current flowing through Q1 also flows through current-sensing resistor R7. The voltage developed across R7 is applied to IC1 as a control voltage, along with the current-adjust, overvoltage-sense, and current-sensing inputs.

As Q1 turns on and off, the secondary voltage developed across T1 is rectified by CR50. The rectified output of about 14 to 20 V is filtered and is applied to various regulators and switches.

The 5-V regulator Q50 produces an output voltage referred to as Always 5 Volts or A 5 V. This output is applied to IC1 as a control or sensing input, as well as to IC50 and IC51.

The rectified output is also applied to switching transistors Q51 and Q52. The output from Q52 is routed through diode CR53 to the battery contacts. For this particular adapter/charger, the battery contacts are on the rear cover and mate with the Ni-Cad battery. The output from Q51 is routed to the front surface contacts to power the camcorder. As discussed, the adapter/charger can power the camcorder or charge the battery, but not both. As a result, Q51 and Q52 are never on at the same time.

The adapter/charger uses sense switches to determine if the camcorder or battery is physically attached. Sense switch SW2 is located on the rear panel of the adapter/charger and closes if the battery is attached. If SW2 is closed, and the CHARGE mode is selected, IC50 turns Q52 on, applying the charging voltage to the battery contacts. If the battery is not attached, IC50 does not turn Q52 on, and there is no charging voltage applied to the battery contacts. The camcorder sense switch SW3 and Q51 operate in the same way, appying power to the camcorder power contacts only when the battery is not used and the adapter/charger is physically attached to the camcorder.

Samples of the output voltage from IC50 are passed to IC1. The sample applied through CR54 shuts IC1 down if the rectified output voltage exceeds

Sec. 7-2   AC Adapter/Charger   205

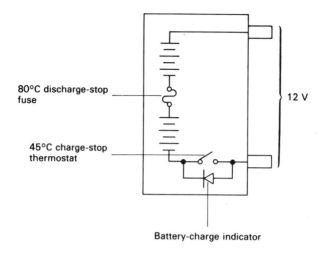

FIGURE 7-3 Internal circuit of typical Ni-Cad battery

a certain level. The sample applied through R61 permits the output current of the adapter/charger to be adjusted for a given load. IC50 also controls the battery discharge indicator (LED or EVF display) drive circuit.

Charge time for the battery is controlled by IC51 in association with IC50. IC51 applies a charge to the battery for 120 minutes (or 160 minutes for some camcorders). After the charge time has elapsed, IC51 turns the output control system off and removes the charging voltage from the battery contacts.

As shown in Fig. 7-3, when the battery is fully charged, the 45°C thermostat opens, and a battery-charge indicator (within the battery) turns on. Simultaneously, the current-sense circuits within IC50 stop the charge cycle and turn output switch Q52 off. Note that the fuse within the battery opens if the current is sufficient to raise the internal temperature to 80°C. This disables the battery, but prevents an explosion that can occur if battery temperature is too high.

7-2.3 Adapter/Charger Troubleshooting

Be very careful when servicing the adapter/charger, especially during disassembly. The circuits are generally very compact and can be damaged unless you follow the disassembly instructions carefully.

To check operation of the adapter/charger when used to power the camcorder, start by closing SW3 (the power-sense switch). Then check the B+ voltage applied the primary of T1. If absent or abnormal, suspect F1, CR1, the line filter (if any), or C6.

Next, check the voltage at the cathode of dual-diode CR50. If absent or abnormal, suspect Q1, T1, IC1, R7, R3, or R4. If the output from CR50 appears to be good, check for B+ at Q50, Q51, Q52, and IC50. If absent or abnormal, suspect L50, C51/C52, or the wiring.

Check the output of regulator Q50 (5 V). If missing, suspect Q50. Finally, check the voltage at the camcorder power terminals (typically about 12.4 V). If absent or abnormal, suspect Q51 or IC50.

If the adapter function appears to be good, check operation of the adapter/charger when used to charge the battery by closing SW2 (the battery-sense switch) and selecting the charge mode. Check the voltage at the battery power terminals. Typically, the voltage should be greater than 14 V. If the voltage is absent or abnormal, suspect Q52, CR53, IC50, or IC51.

7-3 SYSTEM-CONTROL OVERVIEW

Figure 7-4 shows the major system-control functions in block form.

Readers already familiar with VCRs will recognize most of the function shown. Except for the camera-control and battery overdischarge functions, all other system-control functions can be found on virtually all VHS VCRs.

In the camcorder of Fig. 7-4, all circuits are under control of a single 64-pin flatpack microprocessor IC901. When IC901 is instructed to place the camcorder in a particular mode of operation, various control functions are executed. IC901 sends signals to various circuits of the camcorder and receive signals from the circuits.

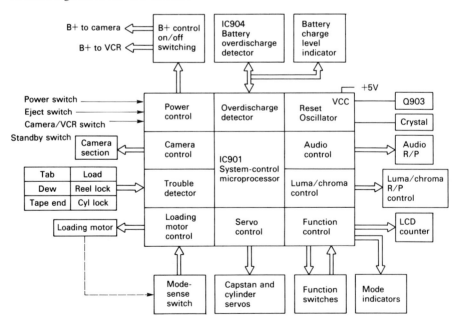

FIGURE 7-4 Major system-control functions

Sec. 7-4   Power-Control Circuits

For example, the loading-motor–control area of IC901 drives the loading-motor circuits. In turn, the circuits operate the motor and various levers and gears to control the tape-transport mechanism mode of operation (Chapter 5). Also associated with the loading-motor system are the mode-sense switches (also called the mechanism-state switches). No matter what name is used, the switches inform IC901 of the tape-transport mechanism position (loading, unloading, unloading stop, etc.). In this way, IC901 maintains virtually automatic control of the camcorder functions.

The following paragraphs describe each of the major functions performed by IC901. Keep in mind that there is reference from the circuit descriptions in Chapters 4, 5, and 6 to this chapter when system control is involved.

7-4   POWER-CONTROL CIRCUITS

Figure 7-5 shows typical power on/off circuits. Figures 7-6 and 7-7 show typical power distribution circuits.

Power is applied through connector PG909-2 (from battery contacts on the rear of the camcorder) and through a switch on the external battery jack J950 to fuse F970. The power is then applied to the emitter of switching transistor Q904 and latching relay RL901. Transistor Q904 is a start-up device used only during initial turn on of the camcorder.

Relay RL901 contains two sets of switches, along with two coils. One coil is powered momentarily to latch the contacts in the on state, while the other coil is used to unlatch the contacts to the off state. A small pulse of energy is required in the appropriate on or off coil to toggle the contacts open or closed for the mode desired. Using this form of latching relay (quite common for camcorders) conserves battery power during operation. (There is no need to draw power continuously to hold the relay closed or open.)

When power switch S056 is pressed, the cathode of one diode within D910 is grounded. This forward-biases D910 and grounds the base of Q904 through ZD901. Transistor Q904 turns on, applying 12 V to the input of 5-V regulator IC905. The 5-V output of IC905 is applied to the B+ input of IC901 at pin 26 and to the reset circuit Q903/ZD902.

With B+ and reset applied, IC901 turns on, and monitors the various input pins for a mode instruction. This initial start-up mode occurs fast enough that IC901 senses a low at pin 12 (due to the power switch S056 being pressed and held). At that time, IC901 realizes that power on is requested and applies a pulse from pin 52, turning on Q901. Transistor Q901 applies a low to the turn-on coil of RL901. The power-on pulse from pin 52 of IC901 occurs for about 100 ms. This is long enough to energize the turn-on coil of RL901, allowing the coil to latch the contacts in the on position.

With RL901 latched on, the 12-V power from RL901 to IC905 is applied

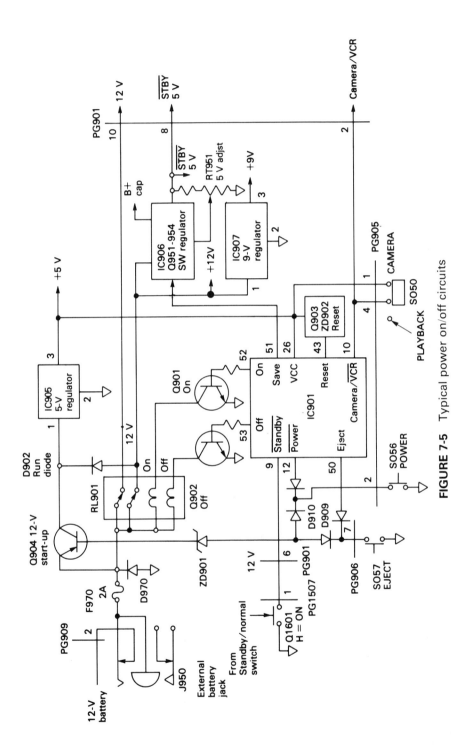

FIGURE 7-5 Typical power on/off circuits

Sec. 7-4　Power-Control Circuits

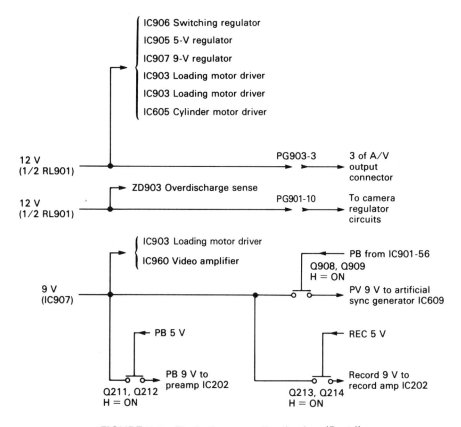

FIGURE 7-6 Typical power distribution (Part I)

through D902, bypassing the start-up transistor Q904. The 12-V power is also applied to regulators IC906 and IC907. Switching regulator IC906 supplies B+ to the capstan motor system and standby 5-V power to the camcorder, IC907 supplies 9-V power to the camcorder.

When the standby/normal switch is set to NORMAL, the power supply provides four voltages: 12 V, 9 V, 5 V, and standby 5 V, as shown in Figs. 7-6 and 7-7. When the standby/normal switch is set to STANDBY, Q1601 is turned on, applying a low or ground to pin 9 of IC901. This causes pin 51 of IC901 to go high, turning off switching regulator IC906. With IC906 off, B+ power is removed from the capstan motor, and standby 5-V power is removed from various components. This saves battery power, but keeps the camcorder circuits ready for instant operation.

When eject switch S057 is pressed, Q904 is turned on, applying start-up B+ to IC905. This turns IC901 on through pins 26 and 43. IC901 senses that S057 is closed (through pin 50 of IC901) and applies a power-on pulse to Q901, setting RL901 to on. IC901 activates the loading motor (Fig. 7-6)

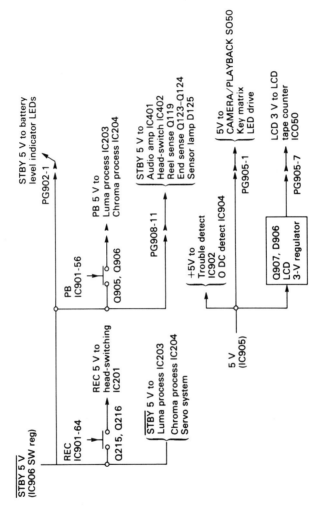

FIGURE 7-7 Typical power distribution (Part II)

to eject the cassette (Chapter 5 and Sec. 7-10). After eject operation is complete, IC901 turns the camcorder off by applying a momentary pulse to Q902.

When camera/playback switch S050 is set to CAMERA, 5-V power is applied to pin 10 of IC901. In turn, IC901 places the camcorder in the record-pause mode. When S050 is set to PLAYBACK, pin 10 of IC901 goes low. Under these conditions, the VCR portion of the camcorder operates as a conventional VCR, with the EVF as a monitor. Also, an optional external monitor can be connected through the A/V output connector or through an optional RF adapter (modulator).

7-4.1 Power-Control Troubleshooting

If the camcorder does not turn on with S056 pressed, substitute a known-good battery or adapter/charger. Next, check fuse F970. If the fuse is open, and blows a second time, suspect D970, IC905, IC906, IC907, or the various motor-drive ICs (Chapter 5).

If 12-V power is present on both sides of F970, check for B+ at pin 1 of IC905 when S056 is pressed. If missing, suspect S056, D910, ZD901, and Q904.

If 12-V power is present at pin 1 of IC905, check for 5-V power at pin 3 of IC905. If missing, suspect IC905.

If there is 5-V power at pin 3 of IC901, check for a turn-on pulse at the base of Q901 (with S056 pressed). If missing, suspect a defective reset circuit, Q903/ZD902, or IC901. If the pulse is present, but there is no 12-V power at the anode of D902, suspect Q901 or RL901. It is also possible that Q902 is leaking and prevents RL901 from moving the contacts to the on (closed) state.

If 12-V is present at the anode of D902, check for standby 5-V power at pin 8 of connector PG901. If missing, suspect IC906 and/or the associated transistors Q951–Q954.

Also check the logic at pin 51 of IC901. If low (NORMAL), all the power circuits should be operative. If high (STANDBY), check that the standby/normal switch is in NORMAL. If normal operation is selected, Q1601 should be off, and pin 9 of IC901 should be high.

If 12-V is present at the anode of D902, check for 9-V power at pin 3 of IC907. If missing, suspect IC907.

7-5 BATTERY OVERDISCHARGE DETECTION CIRCUITS

Figures 7-8 and 7-9 show typical battery overdischarge detection (ODC) circuits. The circuits of Fig. 7-8 use three LEDs to show battery condition or

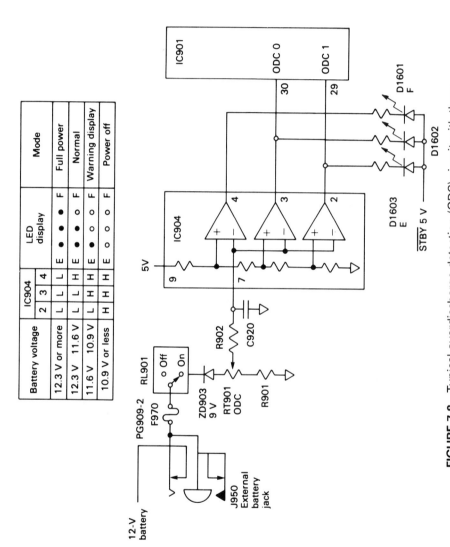

FIGURE 7-8 Typical overdischarge detection (ODC) circuit with three LEDS

Sec. 7-5   Battery Overdischarge Detection Circuits   213

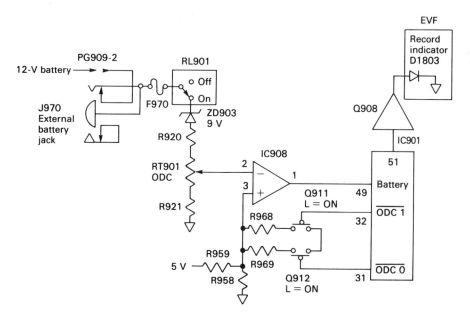

FIGURE 7-9 Typical overdischarge detection (ODC) circuit with EVF display indicator

charge life, while the circuits of Fig. 7-9 display characters in the EVF to indicate battery condition.

If the battery voltage is 12.3 V or more, all three LEDS are on, or the EVF display is E----F.

If the battery voltage is between 12.3 and 11.6 V, the first two LEDs (from the left) are on, and the third LED is off, or the EVF shows E--.

If the battery voltage is between 11.6 and 10.9 V, one LED is on, or the record LED in the EVF blinks at 1 Hz, and the EVF display shows E-. (This indicates that there is about 3 to 5 min of record time remaining.)

When the battery voltage drops below 10.9 V, all three LEDs turn off, there is no EVF display, and the camcorder is placed in the power-off mode by IC901. (However, the cassette can still be ejected by the power remaining in the battery.)

In the circuit of Fig. 7-8, 12-V power is routed through F970 and the contacts of RL901 to the cathode of ZD903. The 12-V power is reduced and appears across ODC control RT901. The reduced voltage from RT901 is applied to pin 7 of IC904, a three-level voltage comparator. The three outputs (at pins 2, 3, and 4) from IC904 indicate the level of the voltage applied to pin 7. The three outputs drive the corresponding LEDs to indicate the battery voltage.

The outputs from pins 2 and 3 of IC904 are also applied to pins 29 and

30 of IC901. If both outputs are high, indicating that the battery voltage is less than 10.9 V, IC901 turns the camcorder off.

In the circuit of Fig. 7-9, 12-V power from the battery is routed through F970, RL901, ZD903, and ODC control RT901 to the inverting input of a comparator within IC908. The reference voltage applied to the noninverting input is taken from the 5 V supplied from D/A converter Q911/Q912. Both Q912 and Q911 are turned on and off by ODC signals from IC901.

When the battery voltage applied to the inverting input of IC908 is lower than the reference voltage applied to the noninverting input, the output goes low, and IC901 detects the battery voltage. IC901 then decides battery condition and produces the corresponding display (as shown by the display truth table on Fig. 7-8).

Note that when the battery voltage is between 11.6 and 10.9 V, the EVF display shows E-, with "-" flashing, and the record LED in the EVF blinks at 1 Hz during record. (If the camcorder is not in record, the record indicator LED is turned off.)

7-5.1 Average Battery-Charge Life

Figure 7-10 shows the average charge life of a typical camcorder battery. Note that the maximum operating time for some batteries is 1 hr, while other batteries provide 2 hr.

No matter what the maximum operating time, the output voltage of a typical Ni-Cad battery drops within the first few minutes of operation. Output voltage drops from full-charge to a plateau where the voltage remains relatively constant for the maximum charge life. Near the end of charge life, the output voltage starts to drop at a rapid rate as shown. This particular characteristic is consistent for most Ni-Cad batteries on the market today.

7-5.2 Battery ODC Troubleshooting

If the camcorder appears to operate normally in all modes, but the ODC circuits are not normal, it is a relatively simple matter to check operation of the circuits in Fig. 7-8 and 7-9. Start by monitoring the voltage at pin 7 of IC904 (Fig. 7-8) or at pin 2 of IC908. Then check the status of pins 2, 3, and 4 of IC904 (Fig. 7-8), or the reference voltage at pin 3 of IC908.

If the record LED D1803 in the EVF does not blink when the battery voltage is between 11.6 and 10.9 V, check for a 1-Hz signal at pin 51 of IC901. If absent, suspect IC901. If present, but D1803 does not blink, suspect Q908 and D1803.

It is possible to check the truth tables shown on Fig. 7-8 by connecting a variable power source to the battery input. Simply vary the input voltage to the value shown on the truth tables, and check that the ODC circuits respond accordingly.

Sec. 7-6   Function Switch Circuits   215

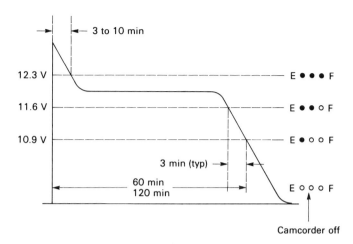

FIGURE 7-10   Average charge life of a typical camcorder battery

If the camcorder of Fig. 7-8 fails to turn on with a known-good battery or other power source, press power switch S056 and check the status at pins 2 and 3 of IC904. If the battery voltage is 12.3 V, or more, pins 2 and 3 (and pin 4) should all be low. If not, suspect IC904, ZD903, or RT901. Also check adjustment of RT901.

Keep in mind that if pins 2 and 3 of IC904 are high, for whatever reason (low battery or a malfunction), IC901 will turn the camcorder off.

7-6 FUNCTION SWITCH CIRCUITS

Figure 7-11 shows typical function switch circuits.

The VCR portion of the camcorder uses a keyboard input matrix system similar to that of most VCRs. The clock pulse outputs, phase 0 and phase 1 from pins 35 and 36 of IC901, are applied to the keyboard switches through diodes D050/D051. Outputs from the keyboard switches are applied to the matrix inputs at pins 18, 19, and 20 of IC901. The matrix system operates when the clock pulses are active low.

The VCR control system also monitors various inputs from the CAMERA mode switches, such as the start/stop switch S1604 connected to pin 48 of IC901. Likewise, an external record remote switch can be used when connected through connector J101 to pin 48 of IC901. A low input to pin 48 from either source toggles the camcorder from record pause to record mode, and vice versa.

Another function input is taken from review switch S1603 to pin 49 of IC901. When S1603 is pressed, IC901 places the VCR in REVERSE PLAYBACK for 4 sec, FORWARD PLAYBACK for 4 sec, and then returns back to the RECORD-PAUSE mode.

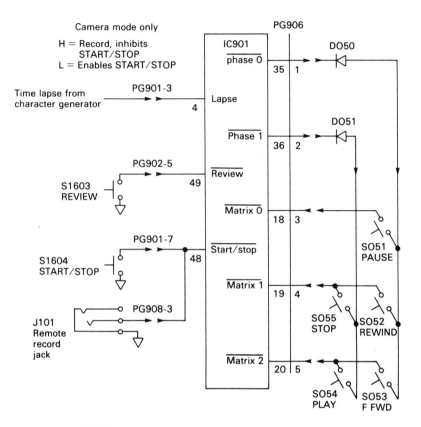

FIGURE 7-11 Typical function switch circuits

An additional input to IC901 is from an optional character generator that controls the record period of the camcorder through the input at pin 4 of IC901. When pin 4 is low, the system operates in the normal manner under control of S1604. When pin 4 is made high by a time-lapse signal from the external character generator, the S1604 input at pin 48 is inhibited. After the preprogrammed record period ends, the character generator pulls pin 4 of IC901 low and holds pin 4 low during the wait period of recording. At this time, start/stop switch S1604 is enabled, if needed.

7-6.1 Function Switch Troubleshooting

The first step in troubleshooting the function switch circuits is to check for proper clock pulses at pins 35 and 36. Then check that the clock pulses reach only the correct matrix inputs at pins 18, 19, and 20 when the keyboard switches are pressed. For example, the phase 0 clock pulses should be applied to pin 18 when pause switch S051 is pressed. There should be no other pulses at pin 18 and no pulses at pins 19 and 20.

If the clock pulses are absent, suspect IC901. The same is true if the correct pulses reach the matrix inputs. If the clock pulses do not reach the correct inputs, suspect the corresponding switch and diodes D050/D051.

Next, check for the correct inputs at pins 4, 48, and 49 of IC901 when the corresponding switch is pressed (or the character generator is operated in the time-lapse mode). If the correct inputs are not available, trace the wiring back to the switches and character generator. If the inputs are available, but the camcorder does not respond, suspect IC901.

7-7 TROUBLE SENSOR CIRCUITS

Figure 7-12 shows typical trouble sensor circuits. Readers familiar with VCRs will recognize most of the trouble sensor functions. However, operation of the trouble sensor circuits is not necessarily the same as found in VCRs.

7-7.1 Dew Sensor

The dew sensor senses moisture in the VCR portion of the camcorder. As moisture increases, pin 9 of IC902 goes high, as does pin 14 of IC902 and pin 14 of IC901. This places the system in the STOP mode. On some camcorders, the power LED blinks to indicate an excess moisture condition. On other camcorders, the word DEW appears in the EVF. When the excess moisture is gone, the voltage at pin 9 of IC902 drops below that at pin 8. This causes pins 14 of IC902 and IC901 to go low and returns the system to normal (usually to the RECORD-PAUSE mode).

7-7.2 End Sensor

Transistor Q101 is turned on for about 0.1 ms, every 3.0 ms, by pulses from pin 63 of IC901. When Q101 is on, the IR end sensor lamp D125 is turned on momentarily. Unlike most VCRs, D125 is not operated continuously in a camcorder to minimize power consumption.

When the tape reaches the end, while moving in the forward direction (play, fast forward, forward search), the clear leader on the tape end allows the IR light from D125 to fall on the supply end sensor Q123. This causes pins 5 and 2 of IC902, and pin 27 of IC901, to go high, placing the system in the STOP mode.

When the tape reaches the end in the reverse direction, the light from D125 passes through the clear leader to takeup end sensor Q124. This causes

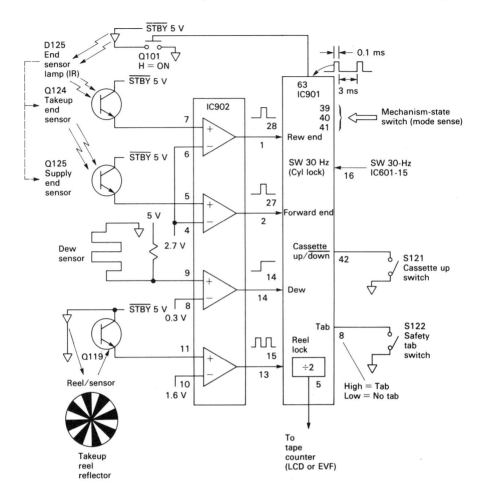

FIGURE 7-12 Typical trouble sensor circuits

pins 7 and 1 of IC902 and pin 28 of IC901 to go high, again placing the system in the STOP mode.

7-7.3 Reel Sensor (or Reel-Rotation Detector)

The reel sensor (or reel-rotation detector) consists essentially of a combined LED and light sensor transistor in one package Q119 and a reflector disc with eight reflective and nonreflective areas located on the underside of the takeup reel.

As the reel rotates, squarewave pulses are generated by Q119. (The

Sec. 7-7   Trouble Sensor Circuits   **219**

light from the LED in Q119 is applied to the disc surface. White segments of the disc reflect light, while black segments absorb light. This produces pulses of light applied to the Q119 light sensor transistor.) The voltage pulses generated by Q119 are applied to pin 15 of IC901 through IC902.

If the reel stops rotating (for 1.5 sec in FF or rewind, or 3 sec in other modes), the pulses are removed, and IC901 places the camcorder in stop mode. During normal operation, while the reel is rotating, the pulses are divided by IC901 and used to feed the tape counter (as described in Sec. 7-8).

7-7.4 Cassette-Up Switch

The cassette-up switch S121 is connected to pin 42 of IC901. When a cassette is loaded into the holder and the holder is moved to the down position (Chapter 5), pin 42 is low. If the cassette is up (for any reason), pin 42 goes high, and IC901 places the camcorder in stop.

7-7.5 Safety Tab Switch

When a cassette without a safety tab (not to be recorded) is installed, safety tab switch S122 is closed, making pin 8 of IC901 low. This causes IC901 to place the camcorder in stop and prevents the system from entering the RECORD or RECORD-PAUSE modes. (Pins 59 and 64 of IC901 remain low.)

7-7.6 Mechanism-State (Mode-Sense) Switches

The mechanism-state or mode-sense switches are also a form of trouble sensor on most camcorders. These switches apply signals to IC901 at pins 39 through 41. One signal is applied at the end of unloading operation to place the camcorder in stop. Other signals from the switches to IC901 are used to determine if the state of the VCR mechanism agrees with the mode selected by the function switches (Sec. 7-6). If the two do not agree, IC901 places the camcorder in stop. The relationship between the mechanism and the state or sense switches is discussed further in Sec. 7-10.

7-7.7 Cylinder-Lock Detector

The cylinder-lock signal is the SW 30-Hz signal used by the cylinder servo (Chapter 5). This signal is applied to pin 16 of IC901, which detects the pulse width. If the pulse width is less than a certain value, indicating that the cylinder motor has slowed down (or stopped), IC901 places the camcorder in stop.

7-7.8 Trouble Sensor Troubleshooting

If the camcorder does not go into play, or record, or any particular mode, check the inputs of all trouble sensors at the corresponding pin of IC901. If the signals are absent or abnormal, trace the signal back to the source (sensor or switch). If all signals are normal, suspect IC901.

Make certain that *both* pins 27 and 28 of IC901 are low. If either pin is high, the tape may be at either end (and can go no farther!).

If there appears to be a problem with the tape end sensors, check for pulses at pin 63 of IC901. Also check that the cathode of D125 is pulsed by Q101. You cannot see the IR light from D125, but the cathode of D125 should be pulsed between high and ground.

If the VCR mechanism loads tape from the cassette, and then immediately unloads tape and/or stops, look for pulses at pin 15 of IC901 at the same time when the tape just starts to move (takeup reel rotating). If the pulses are missing, suspect Q119, IC902, or servo problems (Chapter 5).

Note that if the VCR does not make any attempt to load tape from a cassette (after the cassette is pulled in and down by the load mechanism), the reel-rotation detector and cylinder-lock detector are *probably not at fault*. These two trouble sensors place the VCR in stop only after the cassette is loaded, the tape is loaded from the cassette, and the tape starts to move. If the VCR does not load tape, the dew sensor, cassette up/down switch, and the mechanism switches are far more likely suspects.

7-8 TAPE COUNTER CIRCUITS

Figures 7-13 and 7-14 show typical counter circuits. The circuits of Fig. 7-13 use an LCD display. The circuits of Fig. 7-14 show tape counter information on the EVF (along with the battery level indication, as discussed in Sec. 7-5).

The LCD counter of Fig. 7-13 is contained in a single module IC050. When power is first applied, reset transistor Q051 erases the memory in IC050 and initializes the display to 0000. Power for IC050 is taken from a 3-V source through Q907. Capacitor C911 provides a backup for the LCD display when the 3-V source is removed (when the VCR is turned off or in the event of a momentary power loss). The charge on C911 keeps the LCD display on for about 3 hr in the normal operating mode.

When the save input at pin 4 of IC050 is pulled low by IC901, IC050 is placed in a POWER-SAVE mode to reduce power consumption. This power-save feature, along with the charge on C911, allows the LCD display to show and retain the last tape-count information for several days (with all power off).

The LCD display is driven by reel pulses from pin 5 of IC901 to pin 1

Sec. 7-8 Tape Counter Circuits

Camcorder Mode	Input to ICO50		Counter Mode
	Counter 1 (pin 10)	Counter 0 (pin 9)	
Stop	1	1	Stop
Play, fast forward	1	0	Increment count
Rewind	0	1	Decrement count

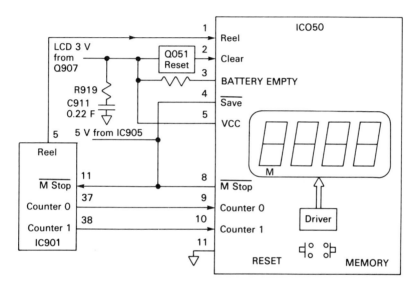

FIGURE 7-13 Typical tape counter with LCD display

of IC050. IC901 divides down the reel-rotation pulses from the reel sensor by a factor of two, as described in Sec. 7-7.

The LCD display also receives two control inputs at pins 37 and 38 from IC901. These inputs control the mode of operation (count-up, count-down, memory-on, etc.) as shown by the truth table on Fig. 7-13.

The LCD display generates a memory-stop output signal at pin 8. This signal is applied to IC901 at pin 11. If the memory system is turned on (by the memory button on the LCD module), IC050 generates a low (memory-stop) when the count decrements to 0000. This informs IC901 to place the VCR in the stop mode.

The reset button on the LCD module resets the counter to 0000 (as does Q051 when power is first applied). The memory button on the LCD module controls the memory-stop function. The M indicator (below the four-digit counter display) turns on when the memory-stop function is on. If the mem-

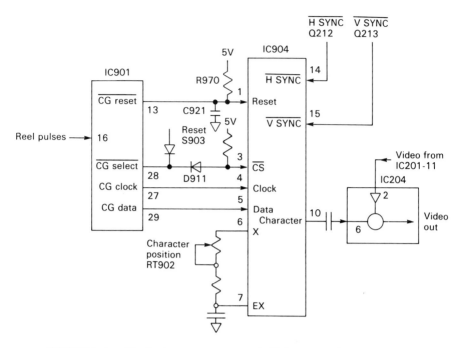

FIGURE 7-14 Typical tape counter with EVF display (character generator)

ory button is pressed when the M indicator is on, the memory-stop function goes off.

The LCD module also has a battery empty indication. This turns on when the battery voltage (applied to pin 3 of IC050) is removed.

In the circuit of Fig. 7-14, character generator IC904 receives character information from IC901, along with H-sync and V-sync information from the signal-processing circuits (Chapter 6).

IC904 generates a character signal that is synchronized with the horizontal and vertical sync signals. The character signal at pin 10 of IC904 is mixed in IC204 with the video signal applied from the signal-processing circuits. Character-position control RT902 determines the position or location of the characters on the EVF display.

7-8.1 Tape Counter Troubleshooting

The first step in troubleshooting the tape counter circuits of Fig. 7-13 is to check all the inputs at IC050. If all the inputs (including the power inputs) are correct, but the LCD display is absent or abnormal, suspect IC050. For example, if the VCR is in play or fast forward (pin 9 low, pin 10 high), but the counter does not increment, suspect IC050.

Sec. 7-9   Mode Indicator Circuits

If any one of the inputs is absent or abnormal, trace the line back to the signal source. For example, if the clear signal is not applied to IC050 (with 3-V power available), suspect Q051. As another example, pin 1 of IC050 should show reel pulses (with the tape moving in either direction). If not, suspect IC901 or the reel sensor.

Keep in mind that if the tape is moving in either direction, the reel-rotation pulses must be present in IC901. If not, the trouble sensor system will place the VCR in stop.

The first step in troubleshooting the circuits of Fig. 7-14 is to monitor all the inputs to the character generator (pins 1, 3, 4, 5, 14, and 15 of IC904), as well as the output at pin 10. If any of the inputs are absent or abnormal, trace the line back to the source. If the inputs are good, but the output is absent or abnormal, suspect IC904. While it is not practical to interpret or decode the codes (particularly on pin 5 of IC904 from pin 29 of IC901), the presence of pulses on the line usually indicates that the function is normal.

Also try adjusting RT902 as described in Chapter 8. It is possible that the characters are off screen, due to a severe misadjustment of RT902.

7-9   MODE INDICATOR CIRCUITS

Figure 7-15 shows typical mode indicator circuits.

In addition to producing signals that determine the operating mode, IC901 also produces signals that control the indicators to show the mode selected. As discussed in Sec. 7-6, the operating mode is controlled by signals applied to IC901 from the function switches. In turn, IC901 provides signals to the mode-indicator LEDs. In the circuit of Fig. 7-15, there are five mode-indicator LEDs D080-D084, in addition to the record/low-battery indicator D1803 in the EVF.

The mode-indicator drive signals from IC901 are applied to D080-D084 through transistors Q052-Q057. The $\overline{\text{phase 0}}$ and $\overline{\text{phase 1}}$ signals from pins 35 and 36 of IC901 turn on transistors Q052 and Q053, respectively. When Q052 is on (phase 0 low), 5-V power is applied to the anodes of D080-D083. When Q053 is on (phase 1 low), 5-V power is applied to the anode of D084.

When transistors Q054-Q057 receive drive signals (high) from IC901, the cathode of the corresponding LED goes low. The LED turns on, if the anode is connected to 5-V power. For example, if phase 0 (pin 35) is low and pin 34 is high, D080 turns on. Note that in search operation, either D082 or D083 is pulsed at a 1-Hz rate by IC901 (at pins 32 and 33).

During record operation, pin 6 of IC901 goes high. This turns Q1513 on and applies 5-V power to D1803 in the EVF. If the battery voltage drops below 11.6 V during record, pin 6 of IC901 goes low, and the output at pin 7 of IC901 is pulsed, causing Q1513 and D1803 to pulse on and off. When D1803 starts to blink, this indicates that the battery voltage is low, and about 3 to 5 min of record time are left.

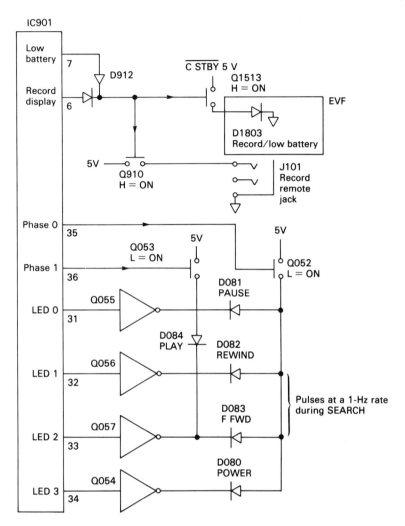

FIGURE 7-15 Typical mode indicator circuits

The outputs at pins 6 and 7 of IC901 are applied to Q910 through D912. Transistor Q910 applies 5-V power to an optional remote start/stop switch through J101. This provides power for an LED on the remote switch.

7-9.1 Mode Indicator Troubleshooting

The first step in troubleshooting the mode indicator circuits is to check if the camcorder is performing the function. If the function is correct, but the mode is not being indicated properly, the problem is in the circuits of

Sec. 7-10 Loading Motor Circuits

Fig. 7-15. On the other hand, if the function is not correct, the problem is elsewhere, probably in the function switch input circuits (Sec. 7-6).

For example, if rewind switch S052 (Fig. 7-11) is pressed, and the camcorder does not rewind, the problem is in the function switch circuits, probably S052 or IC901. If the camcorder does rewind, but D082 (Fig. 7-15) does not turn on, the problem is in the mode indicator circuits, probably D082, Q052, Q056, or IC901.

This same troubleshooting approach can be applied to all remaining circuits shown on Fig. 7-15.

7-10 LOADING MOTOR CIRCUITS

Figure 7-16 shows typical loading motor circuits.

During VCR operation, the mechanism must be placed into different modes of operation for play, record, rewind, search, and so on. All these functions are done with a single loading motor. As discussed in Chapter 5, the loading motor loads the tape, ejects the cassette basket, releases the brakes, engages the fast forward/rewind gear assembly, and engages the playback assembly. It is essential that you study the descriptions of mechanical operation (Sec. 5-2) in conjunction with the following.

At the same time the loading motor is driven, the mechanism-state switch (also called the mode-sense switch) moves to tell IC901 the position or state of the tape transport. Control signals from IC901 are applied to the loading motor through IC903. These control signals tell IC903 to power the loading motor in the correct direction and speed. Signals from the mechanism-state switch are applied directly to IC901, at pins 39, 40, and 41. These signals tell IC901 exactly what mode (position) the mechanism is in. During a VCR function, such as play, record, search, and so on, if the proper state is not indicated by these three inputs, IC901 places the VCR in stop (unloads the tape).

The truth table for the inputs and outputs of IC903, shown on Fig. 7-16, indicates the status of the pins on IC903 for each mode. Note that in stop, pins 3 and 7 are floating, and no power or ground is applied to the loading motor. In loading, pin 3 is connected to power, while pin 7 is connected to ground. This turns the motor in the loading direction. In unloading, pin 3 is grounded, while pin 7 is connected to power. This turns the motor in the unloading direction. In BRAKE mode, both lines to the loading motor are shorted to ground. This dissipates the collapsing energy in the loading motor windings, causing the motor to stop immediately.

The truth table for the mechanism-state inputs to IC901, in Fig. 7-16, shows the status of the pins on IC901 for each mode. Note that in the EJECT mode, if power is not applied to the loading motor system, ejecting the tape is not possible unless the case is removed and the hold-latch physically released. Keep in mind that the mechanism-state inputs to IC901 do not control IC901 directly, but tell IC901 the actual status of the mechanism.

7-10.1 Loading Motor Troubleshooting

The first step in troubleshooting a problem in the loading motor circuits is to check that the inputs to IC903 from IC901 are correct and that the outputs from IC903 to the loading motor are correct, using the IC903 truth table.

If the inputs from IC901 to IC903 are not correct, suspect IC901 or the function switches (Sec. 7-6).

IC903 IN		IC903 OUT		
Pin 4	Pin 6	Pin 7	Pin 3	Mode
0	0	—	—	Stop
0	1	0	1	Loading
1	0	1	0	Unloading
1	1	0	0	Brake

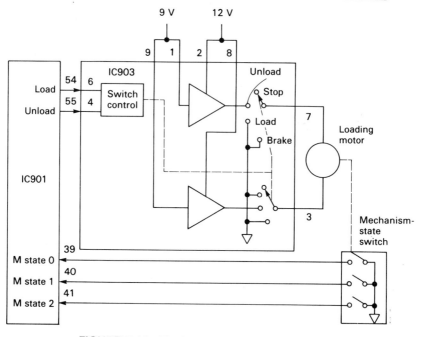

FIGURE 7-16 Typical loading motor circuits

Sec. 7-11 Capstan Motor Circuits

Mechanism state switch position	IC901 Inputs			Mode
	M state 2 (pin 41)	M state 1 (pin 40)	M state 0 (pin 39)	
Eject	0	1	1	Eject
Unloading, stop	0	1	0	—
Fast forward/rewind	1	1	0	Fast forward/rewind
Loading/unloading	1	0	0	—
Record-pause	1	0	1	Record-pause
Play/record	0	0	1	Record/play/play-pause/search

FIGURE 7-16 (continued)

If the inputs to IC903 are good, but the outputs from IC903 are not good, suspect IC903. Likewise, if inputs to the loading motor are correct, but the loading motor does not respond properly, suspect the motor.

Next, check the status of the mechanism-state inputs to IC901, using the IC901 truth table. Compare this to the descriptions given in Sec. 5-2. For example, if the VCR is in stop, and the play switch is pressed, the loading motor starts to rotate in the loading direction (pins 3 and 6 of IC903 high) as discussed in Sec. 5-2.6. When the mechanism is in stop, pin 40 of IC901 should be high, with pins 39 and 41 low. When the mechanism goes into the PLAY mode, pin 39 of IC901 should go high, with pins 40 and 41 low.

Remember, during loading operations, if load completion is not indicated by the mechanism-state switch, IC901 reverses direction of the loading motor and unloads the tape into the cassette. Also remember that if IC901 does not provide the correct outputs to IC903, it is possible that one of the trouble sensors (Sec. 7-7) may be the problem.

7-11 CAPSTAN MOTOR CIRCUITS

Figure 7-17 shows typical capstan motor-control circuits.

During various modes of operation, IC901 controls both the direction and speed of the capstan motor. As discussed in Sec. 5-2, the capstan motor is coupled to various mechanical assemblies through belts and gears, thus supplying tape movement in PLAY, RECORD, REWIND, FAST FORWARD, and SEARCH modes. For these different modes of operation, the tape speed varies slightly. As a result, a different amount of voltage must

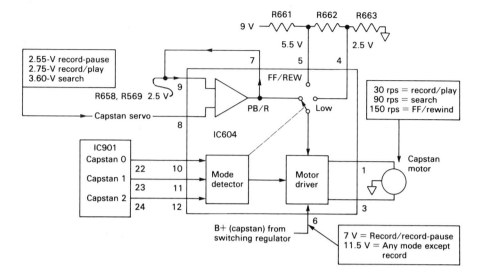

IC604 Input				
Pin 12	Pin 11	Pin 10	Capstan motor	Mode
0	0	0	Stop	Stop/pause/loading
0	0	1	Play (forward)	Record/play/forward-search
0	1	0	Play (reverse)	Reverse-search
0	1	1	Fast forward (fast)	Fast forward
1	0	0	Rewind (fast)	Rewind/unload
1	0	1	Fast forward (slow)	Fast forward start
1	1	0	Rewind (slow)	Slack removal at unload end
1	1	1	Stop	Brake

FIGURE 7-17. Typical capstan motor-control circuits

Sec. 7-12 Mode Control Circuits

be applied to the capstan motor in each mode. This is controlled by IC604 in response to signals from IC901, as shown by the truth table of Fig. 7-17.

A switchable B+ voltage from switching regulator IC906 is applied to pin 6 of IC604. This B+ is switched between two voltage levels, depending on the mode of operation. During record, B+ to IC604 is dropped to about 7 V. This reduces power dissipation in IC604 and improves efficiency. During any other mode (PLAY, SEARCH, REWIND, FAST FORWARD), 11.5 V is applied to IC604.

During PLAY and RECORD, the capstan servo applies a control voltage to pin 8 of IC604. The control voltage is amplified to the motor-drive circuits within IC604. In turn, drive signals from IC604, pins 1 and 3, are routed to the capstan motor. As discussed in Sec. 5-5, feedback signals are applied to the servo system when the capstan motor rotates. These feedback signals indicate capstan speed. The servo circuits apply the correct voltage to IC604 (2.75 V in the case of play or record) to keep the capstan motor speed at 30 rps.

During REVERSE SEARCH, the mode-detect circuits in IC604 are instructed by the three control signals from IC901 (capstan 0, 1, and 2) to change polarity of the drive voltage to the capstan motor. The control voltage is still applied by the servo system to maintain proper speed and phase control of the capstan motor. The servo circuits apply the correct voltage to IC604 (3.6 V in the case of search) to keep the motor speed at 90 rps. In forward search, the capstan motor still runs at 90 rps, but in the forward direction.

During FAST FORWARD and REWIND, the mode-detect circuits in IC604 are instructed by IC901 to select the voltage at pin 5 of IC604 as the control voltage (instead of the voltage from the servo). The voltage at pin 5 of IC604, derived from the voltage divider R661/R662/R663, is about 5.5 V. This increases the capstan motor speed to 150 rps.

As shown by the truth table of Fig. 7-17, the capstan motor is slowed down at the start of fast forward and at the unloading end of rewind (to remove tape slack). This is done by selecting 2.5 V at pin 4 of IC604 and is an automatic function IC901.

7-11.1 Capstan Motor Troubleshooting

The first step in troubleshooting a problem in capstan motor circuits is to operate the capstan motor in the three basic modes (RECORD/PLAY, SEARCH, FAST FORWARD/REWIND) and check capstan rotation.

If the capstan motor fails to rotate in any mode, suspect the capstan motor or IC604. Check if voltages are applied to the capstan motor from pins 1 and 3 of IC604. If so, the motor is probably at fault. If there is no voltage from IC604 to the motor, IC604 is probably at fault.

If the capstan motor rotates, but at the incorrect speed, in the wrong direction, or not in all modes, check all inputs to IC604, pins 4 through 12. If any of the inputs are absent or abnormal, track back to the source.

Pay particular attention to the inputs at pins 10, 11, and 12 of IC604. Make certain that the correct inputs are applied from IC901, as shown by the truth table on Fig. 7-17.

Note that in PLAY and RECORD, the speed of the capstan motor is determined by the voltage applied at pin 8 of IC604 (from the capstan servo, Chapter 5). So, if operation is normal during SEARCH and FAST FORWARD/REWIND, but not in PLAY/RECORD, the problem is likely in the capstan servo.

If the problem occurs only during RECORD, check the voltage at pin 6 of IC604. The voltage should be 7 V for RECORD, and 11.5 V for all other modes.

If the problem occurs only during FAST FORWARD or REWIND, check the voltage at pins 4 and 5 of IC604. The voltage should be 2.5 V at pin 4, and 5.5 V at pin 5.

7-12 MODE CONTROL CIRCUITS

Figure 7-18 shows typical mode-control circuits.

When the system-control microprocessor IC901 is instructed to place the camcorder in a particular mode, various control functions are executed as previously discussed. Various control lines from IC901 are turned on (generally active high) to activate the corresponding circuit. The essential outputs from IC901, for a particular mode of operation, can be checked by reference to Fig. 7-18 and to the following descriptions. (If you are very fortunate, you may find similar information in the service literature for the camcorder you are servicing. Do not count on it!)

The pause line at pin 1 is supplied to the servo circuit. When IC901 is in pause, pin 1 goes high. If the camcorder is in STILL, the high at pin 1 turns on the artificial vertical pulses for vertical jitter correction. If the camcorder is in record, the high turns on the phase-match edit function and inhibits the $\frac{1}{2}$-V SYNC signal. As discussed in Chapter 5, the phase-match edit function operates when the camcorder is switched from RECORD to PAUSE.

The search line at pin 2 is also supplied to the servo circuit and goes high when IC901 is in SEARCH. The high at pin 2 controls the cylinder/capstan phase and speed circuits (to operate at a fixed speed) and generates artificial vertical pulses.

The ASBL (assemble) line at pin 3 is also supplied to the servo circuit and goes high when IC901 is in the CAMERA mode, and record is entered from RECORD/PAUSE. The high at pin 3 controls the phase-match edit function.

Sec. 7-12   Mode Control Circuits   231

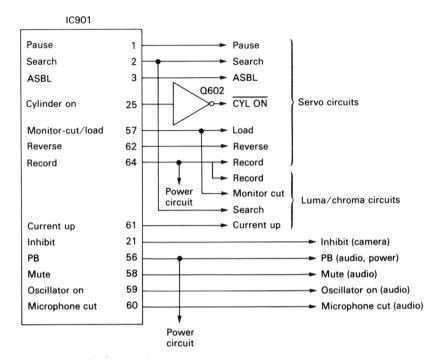

FIGURE 7-18 Typical mode-control circuits

The cylinder-on line at pin 25 is also supplied to the servo circuit and goes high when IC901 is in RECORD, PLAY, VISUAL SEARCH, or STILL modes. This turns on the cylinder/capstan phase and speed circuits. Note that the high at pin 25 is inverted to a low by Q602.

The monitor-cut/load line at pin 57 is supplied to both the servo and luma/chroma circuits and goes high when IC901 is in LOAD. This inhibits the recording of video signals, and the output of playback video signals, during load.

The reverse line at pin 62 is supplied to the servo circuit and goes high when IC901 is in reverse. This reverses the cylinder speed circuit.

The record line at pin 64 is supplied to the servo, luma/chroma, and power circuits and goes high when IC901 is in RECORD. This switches the related circuits to RECORD.

The oscillator-on line at pin 59 is supplied to the audio circuits, and also goes high when IC901 is in RECORD. This switches the audio circuits to record and turns on the audio oscillator circuit. Note that pins 59, 25, and 64 must be on to record.

The current-up line at pin 61 is supplied to the luma/chroma circuits and goes high when IC901 switches from RECORD/PAUSE to RECORD. This increases the video recording current for phase-match edit and occurs at the same time pin 3 goes high.

The inhibit line at pin 21 is supplied to the camera circuits and goes high when IC901 switches to CAMERA mode. The high lasts for 10 sec when power is on and the CAMERA mode is first selected or for 1 sec when recovering from a STANDBY mode. Either way, the camera video signals are inhibited for a brief period.

The playback line at pin 56 is supplied to the audio and power circuits and goes high when IC901 is in PLAYBACK. This switches the related circuits to PLAY. Note that pin 25 must also be on for playback.

The mute line at pin 58 is supplied to the audio circuits and is high in all modes except RECORD and PLAYBACK. This inhibits the RECORD and PLAYBACK audio signals in all other modes.

The microphone-cut line at pin 60 is supplied to the audio circuits and is high when IC901 is in PLAYBACK or POWER-SAVE modes. This inhibits the microphone output during these modes.

7-12.1 Mode-Control Troubleshooting

It is difficult at best to locate quickly a malfunction in the mode-control circuits of any camcorder. This is because there are so many output control lines from IC901 to various circuits in the camcorder. A failure on any one control line can affect more than one operating mode. Likewise, most operating modes depend on signals from more than one line (just to make things more difficult).

Fortunately, failure in the mode-control circuits usually shows up as a failure to perform one or more operating functions. So the first step is to set the camcorder controls as necessary to select a particular operating function, and then *check all related outputs* from IC901.

If the outputs are incorrect, suspect IC901 or the function switches (operating controls). If the outputs from IC901 are good, but the function is not performed, then check the corresponding circuit or circuits, using Fig. 7-18 as a starting point.

For example, during playback, pins 25 and 56 must be high. Pin 57 goes high during the play-load period. (Note that play-load occurs when play is first selected, during the time it takes to load the tape on the mechanism).

If there is a steady high at pins 25 and 56 (and a momentary high at pin 57 during LOAD) when PLAY is selected, but the camcorder does not go into PLAY, trace the circuit from pins 25 and 56.

The high from pin 25 turns on the cylinder/capstan circuits through inverter Q602. (The output from Q602 must be low so as not to turn on Q601 in the cylinder/capstan circuits.) If the cylinder/capstan circuits do not respond properly to the low from Q602, troubleshoot the cylinder/capstan circuits as described in Chapter 5.

The high from pin 56 turns on the audio circuits and applies PB 5-V

Sec. 7-12 Mode Control Circuits

and PB 9-V signals to the luma/chroma circuits, so that the audio/video heads switch to the PB or PLAYBACK mode. If the audio and/or chroma/luma circuits do not respond properly to the high from pin 56, troubleshoot the audio/chroma/luma circuits as described in Chapter 6.

Once you are satisfied that the playback circuits are turned on and ready to operate normally, also check that there are no circuits or functions turned on that can inhibit playback operation.

For example, if pin 57 remains high after loading is complete, the video signals can be cut during playback. Likewise, if pin 58 goes high during playback, the audio signals can be muted. If pin 59 goes high during playback, the audio oscillator circuits will be turned on, possibly erasing or damaging the audio/video recorded on the tape. If pin 60 goes high during playback, the microphone output can be recorded on the tape.

Finally, if pins 56 and 64 go high simultaneously (due to a malfunction in IC901 or where both pins are shorted to 5 V), there is no way of telling the results (except that playback would be affected).

ELECTRICAL and MECHANICAL ADJUSTMENTS

This chapter describes typical adjustment procedures for both the electrical and mechanical sections of a camcorder. The camcorders covered are those with circuits such as those described in Chapters 4 through 7, using the test equipment described in Chapter 3.

In most cases, the simplified and block diagrams of Chapters 4 through 7 show the electrical position of the controls and test points involved in adjustment. The specific illustrations are referenced in the following test procedures (by figure number). The procedures are also supplemented with illustrations that show the physical location of mechanical adjustment points, as well as waveforms or signals that should appear at the various electrical test points.

Keep in mind that the procedures described here are for reference and are the only procedures recommended by the manufacturer for that particular model of camcorder. Other manufacturers may recommend more or less adjustment. It is your job to use the correct procedures for each camcorder you are servicing.

Also remember that some disassembly and reassembly may be required to reach test and/or adjustment points. We do not include any disassembly/reassembly here for two reasons. First, such procedures are unique and can apply to only one model of player. More important, disassembly and reassembly (both electrical and mechanical) are areas where camcorder service literature is generally well written and illustrated. Just make sure that you observe the notes, cautions, and warnings found in the disassembly/reassembly sections of the camcorder service literature.

Sec. 8-1   Mechanical Adjustments   235

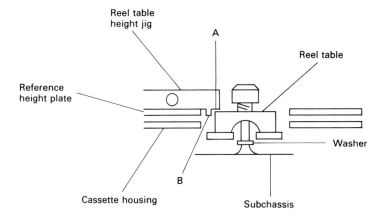

FIGURE 8-1 Reel table height adjustment

8-1 MECHANICAL ADJUSTMENTS

The following paragraphs describe complete adjustment procedures for the mechanical sections of a typical camcorder. Note that the mechanical adjustments are essentially the same for camcorders with tube-type pickups and MOS pickups. The tools and fixtures required for these mechanical adjustments are discussed in Sec. 3-3.

8-1.1 Reel Table Height

Figure 8-1 is the adjustment diagram.

The supply and takeup reel table heights are adjusted by changing the washer stack located under each reel table.

1. Insert the height reference plate into the camcorder.
2. Place the height reference jig on the plate, and check the reel table height. Point A of the jig should pass over the top of the reel table and point B should not.

Note that two sizes of washers are available: 0.25 mm and 0.5 mm. These washers should be used in combination to get equal heights for both reel tables. Check the service literature parts list for washer stock numbers.

8-1.2 Tension Arm Position

Figure 8-2 is the adjustment diagram.

1. Cover the supply tape-end sensor photocell with black tape.
2. Set the tension spring to position B on the spring holder.

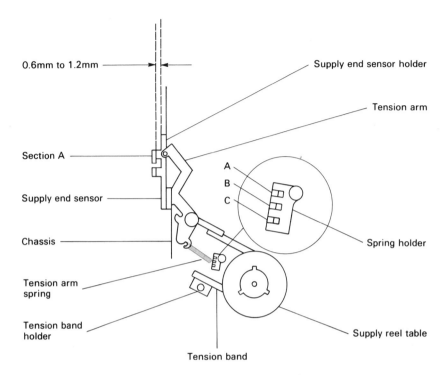

FIGURE 8-2 Tension arm position adjustment

3. Place the cassette holder into down (loaded) position without inserting a cassette tape.
4. Place the camcorder in PLAY mode.
5. When the mechanism has been loaded, loosen the screw holding the tension band holder, and adjust so the clearance between the tension arm and section A of the supply end sensor holder is 0.6 to 1.2 mm.
6. Tighten the screw. Remove the tape.

8-1.3 Back Tension

Figure 8-3 shows the two back-tension gauges. Location of the back-tension spring and related parts is shown in Fig. 8-2. Note that the tension readings are different for each of the two gauges.

When the back tension is properly adjusted, the service test tape (an alignment tape, recorded under laboratory conditions) will play back with minimum *skew error*. (Skew error is picture displacement in the line following head switching.)

1. Place the camcorder in the horizontal position.

Sec. 8-1   Mechanical Adjustments   237

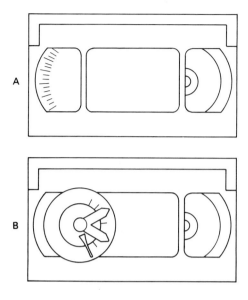

FIGURE 8-3 Typical back-tension gauges

2. Place the camcorder in play mode, with one of the two back tension gauges installed in the cassette holder.
3. The reading should be 27 to 38 on gauge A, or 23 to 32 on gauge B.
4. If the reading is 46 on gauge A, or 33 on gauge B (or higher on either gauge), adjust the tension arm spring in direction A (Fig. 8-2). If the reading is 26 on gauge A, or 22 on gauge B (or lower) adjust the tension arm spring in direction C. This should bring the back tension into tolerance (27–38 gauge A, 23–32 gauge B).

If back tension is still out of tolerance, check the tension arm position as described in Sec. 8-1.2.

8-1.4 Torque Confirmations

Figure 8-4 shows the mechanism components that require torque checks or confirmation. Figure 8-5 shows two torque gauges. The following procedures use the torque gauge and adapter combination. However, some manufacturers recommend the fan-type gauge, particularly for measurement of back tension.

To check brake torque, remove the cassette lid and place the camcorder in the stop mode.

1. Clean the brake surfaces on the reel table using a Kimwipe (or similar applicator) and solvent. *Do not allow solvent to wet the brake pads.*

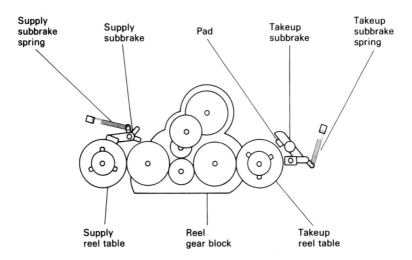

FIGURE 8-4 Mechanism components that require torque checks

2. Attach the adapter to the torque gauge, and place the gauge on the supply reel table.
3. Turn the torque gauge clockwise until the brake begins slipping against the reel table. The torque reading should be more than 140 g/cm.

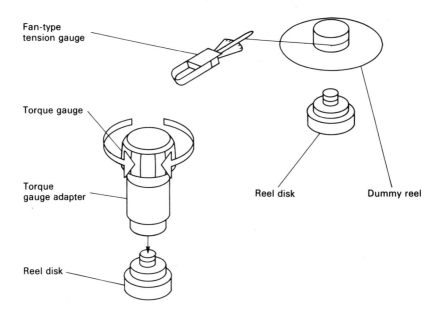

FIGURE 8-5 Typical torque gauges

Sec. 8-1   Mechanical Adjustments   **239**

4. Place the torque gauge on the takeup reel table, and turn the gauge counterclockwise. The torque reading should be more than 80 g/cm.

Note that brake torque problems can cause tape stretch, broken tape, loose tape in cassette, or tape spillage. These symptoms can usually be corrected by proper cleaning. If the symptoms remain after cleaning, replace the main brakes.

To check play, fast forward, and rewind torque, remove the cassette lid.

1. Cover both top end sensors using black tape.
2. Attach the adapter to the torque gauge, and place the gauge on the takeup reel.
3. With the camcorder in PLAY, the torque reading should be 80 to 110 g/cm.
4. With the camcorder in fast forward, the torque reading should be 400 g/cm minimum.
5. Place the torque gauge on the supply reel, and operate the camcorder in rewind. The torque reading should be 400 g/cm.

To check tape slack removal torque, place the torque gauge on the supply reel, and operate the camcorder in play.

1. Press the stop button, and rotate the torque gauge clockwise. The torque reading should be 90 to 200 g/cm.

To check supply reel back torque, place the torque gauge on the supply reel.

1. Press the fast forward (or FF) button, and rotate the torque gauge clockwise. The torque reading should be 4 to 10 g/cm.

As a practical matter, supply reel back torque is difficult to read on a torque gauge. Generally, the indicator will deflect only a slight amount. That is why some manufacturers recommend that fan-type gauge (Fig. 8-5) to measure back tension.

To check takeup reel back torque, place the torque gauge on the takeup reel.

1. Press the rewind button, and rotate the torque gauge counterclockwise. The torque reading should be 4 to 10 g/cm. Again, this may be difficult to measure.

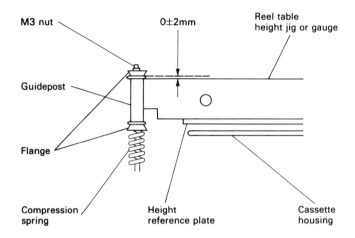

FIGURE 8-6 Tape guidepost height adjustment

8-1.5 Tape Guidepost Height

Figure 8-6 is the adjustment diagram.

1. Install the height reference plate into the camcorder.
2. Place the reel table height gauge on the height reference plate.
3. Adjust the nut on top of the guidepost for a clearance of 0 ± 2 mm.
4. Use the height gauge to make a rough adjustment of the supply guide roller and takeup guide roller. Final adjustment of the guide rollers should be done as described in Sec. 8-1.11.

8-1.6 Impedance Roller Height

Figure 8-7 is the adjustment diagram.

1. Install the height reference plate into the camcorder.
2. Place the reel table height gauge on the height reference plate.
3. Adjust the nut on top of the impedance roller for a clearance of 1.5 to 2.5 mm.

8-1.7 Rough Tape Travel Confirmation

Using a blank tape, place the camcorder in PLAY mode, and check the following. Figure 5-6 shows the location of the components involved as well as the tape path.

Sec. 8-1 Mechanical Adjustments 241

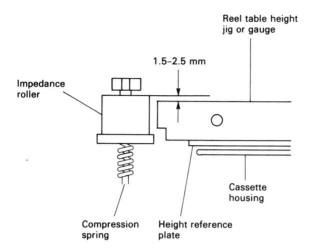

FIGURE 8-7 Impedance roller height adjustment

1. The tape should be in full contact with all tape guideposts or poles.
2. The tape should be crease-free, with no slack.
3. The impedance rollers should be moving freely.
4. The tape should be perpendicular to the longitudinal axis of the heads when crossing the erase and A/C heads.
5. The tape should be centered top to bottom on the head when crossing the full-erase head.
6. The tape should follow the lower edge guide surfaces of the cylinder.

8-1.8 Creasing or Slack Tape Checks

Load the camcorder with a blank tape, and select the PLAY mode. With the tape running, inspect the tape path for creasing or frilling along the top or bottom edges of the tape. (Use a mirror to see the underside of the tape.)

If the tape is not creasing or frilling, check for proper mechanical interchangeability as described in Sec. 8-1.9 and 8-1.10. If the mechanical section passes the interchangeability checks, and is not creasing or frilling, *leave the tape path guide adjustments alone*!

If the tape is creasing or frilling, check the tape as the tape makes contact with the cylinder. The tape should follow the lower edge guide surface on the cylinder. If the tape is high on the guide surface, make a rough adjustment of the guide rollers as discussed in Sec. 8-1.5. Then make a full adjustment as described in Sec. 8-1.11.

8-1.9 Mechanical Interchangeability Considerations

The tape guide adjustments described in Sec. 8-1.11 position the tape so that the prerecorded tracks on the test tape align perfectly with the scan of the video head assembly (on the cylinder). These adjustments ensure that a tape recorded on one VHS camcorder or VCR will play back properly on another VHS camcorder or VCR. Usually, little or no mechanical adjustment is required after routine service (such as replacing the video heads).

So, before you make any tape travel or path adjustments, check interchangeability as described in Sec. 8-1.10. Again, if the mechanical section passes the tests, leave the tape path guide adjustments alone!

If you have done nothing to the mechanical section but replace the video heads, leave all mechanical adjustments alone, but check the PG shifter, record chroma-level, and record luminacle-level adjustments as described in Sec. 8-2.

Of course, if you have been tinkering with the mechanical section in any way, check tape travel as described in Secs. 8-1.7, 8-1.8, and 8-1.10. Unless you are very lucky, you will probably have to adjust the guide rollers as described in Sec. 8-1.12 (and possibly go through every adjustment in this section).

8-1.10 Interchangeability Checks

Figure 6-6 shows the test points involved. Figure 8-8 shows the waveforms.

This confirmation check should be performed after any servicing operation that could adversely affect the tape (motor replacement, tape guide replacement, head replacement, etc.). If the mechanism passes this check, *no tape guide adjustment is needed.*

Note that this check should be made only after the tracking preset adjustment (Sec. 8-2) is made. If the (electrical) tracking preset adjustment is incorrect, it may appear that the tape guide or path is not properly adjusted.

1. Connect the channel 1 scope probe (2 V/div, 2 ms/div) to TP202. Trigger the scope on channel 1.
2. Connect the channel 2 scope probe (50 mV/div) to TP204.
3. Insert a monoscope test tape and set the camcorder to play. Adjust the tracking control for maximum FM envelope on TP204.
4. Adjust the vertical gain on the scope so the maximum envelope is 4 graticule divisions, as shown in Fig. 8-8.
5. Turn the tracking control counterclockwise until the maximum envelope amplitude is 3 graticule divisions.

Sec. 8-1   Mechanical Adjustments   243

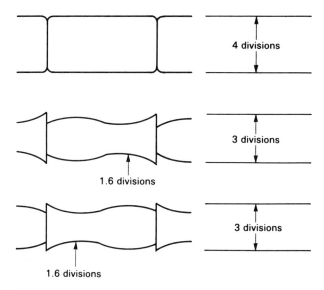

FIGURE 8-8 Interchangeability check waveforms (head amplifier envelope)

6. Check that the minimum envelope amplitude is 1.6 graticule divisions.
7. Turn the tracking control clockwise until the maximum envelope amplitude is 3 graticule divisions.
8. Check that the minimum envelope amplitude is 1.6 graticule divisions.
9. If the mechanical section passes these checks, and there is no tape creasing or frilling, do not make the tape guide roller adjustments.

8-1.11 Supply and Takeup Guide Roller Adjustment

Figures 6-6 and 8-8, used in the interchangeability checks, can also be used in this adjustment, except that scope channel 2 should be set to 100 mv/div.

1. Set the tracking control to the detent or center position and play a monoscope test tape.
2. Loosen the hex screws at the base of the tape guides, and adjust the guide rollers clockwise (down), using a hex screwdriver, until the bottom edge of the tape bows slightly away from the cylinder guide (Figs. 5-1 and 5-6).
3. Adjust the takeup guide roller counterclockwise (up) to get maximum amplitude on the right side of the envelope (Fig. 8-8) at TP204.
   Note that in the event that you cannot get a good head envelope,

it is possible that the A/C head (also known as the ACE head) requires adjustment, as described in Secs. 8-1.12, 8-1.13, and 8-1.14. However, full adjustment of the A/C head is usually not required unless the head has been replaced (or there has been excess tinkering!).

4. Adjust the supply guide roller counterclockwise to get maximum amplitude on the left side of the envelope at TP204.
5. Readjust the tracking control for maximum envelope.
6. Touch up both tape guide rollers for a flat, maximum amplitude envelope, and tighten the hex screws at the bottom of each tape guide.
7. If necessary, adjust the A/C control head horizontal position (Sec. 8-1.14) so the flattest envelope condition occurs at the tracking control detent or center position.

8-1.12 A/C Head Rough Adjustment

Figure 8-9 is the adjustment diagram.

Perform this adjustment before adjusting the A/C head height, tilt, azimuth, and horizontal position. However, it is generally not necessary to adjust the A/C head unless the head has been replaced or there has been extensive work in the tape path. If the head was working properly at one time, and has not been touched, the head does not "go out of line." Of course, if there have been other adjustments in the tape path (guide rollers, impedance rollers, guide posts, etc.), the A/C head should be checked for proper adjustment.

1. Adjust nut A so that the distance between the top surface of nut A and the tip of the threaded shaft is 2.5 to 3.0 mm.
2. Install the height reference plate.
3. Adjust the azimuth screw B, tilt screw C, and screw F so the height difference between the height reference plate and the A/C head base 1 is 1 to 2 mm and the A/C head bases 1 and 2 are parallel.

8-1.13 A/C Head Height/Tilt/Azimuth

Figure 8-9 is the adjustment diagram.

The purpose of this adjustment is to set the A/C head for maximum audio output (indicating that the head is properly aligned with the tape path).

1. Connect the scope (2 V/div, 1 ms/div) to the audio-out jack on the A/V output connector (Fig. 6-8).
2. Play back a 1-kHz audio signal on an alignment tape.
3. Adjust the height nut A and tilt screw C for maximum output.

Sec. 8-1   Mechanical Adjustments   245

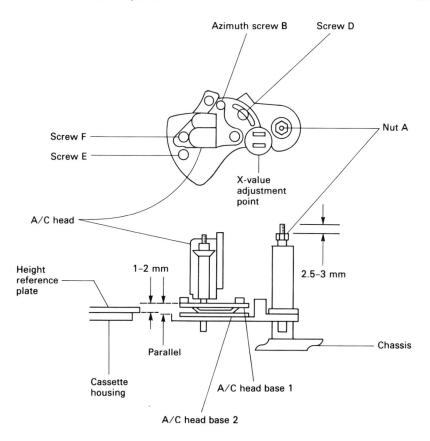

FIGURE 8-9 A/C head adjustment

4. Set the scope time base for 5 ms/div, and play back a 6-kHz (or 7-kHz on some tapes) audio signal on an alignment tape.
5. Adjust the azimuth screw B for maximum output.
6. Repeat steps 2 through 5 for maximum 6-kHz (or 7-kHz) and 1-kHz outputs.
7. Lock the height nut A using lock paint.

8-1.14 A/C Head Horizontal Position

Figure 8-9 is the alignment diagram.

The purpose of this adjustment is to establish proper tape tracking when the tracking control is in the center or detent position. This adjustment should only be made after the tracking preset adjustment (Sec. 8-2) is complete.

1. Connect the channel 1 scope probe (100 mV/div, 5 ms/div) to TP204 (Fig. 6-6).

2. Connect the channel 2 scope probe (1 V/div) to the SW 30-Hz line. Trigger the oscilloscope on channel 2.

Note that with these connections, you are monitoring the video head output but adjusting the control portion of the A/C head for maximum video signal when the tracking control is centered (indicating that tape tracking is correct).

3. Set the tracking control to the detent or center position.
4. Loosen screws D and E.

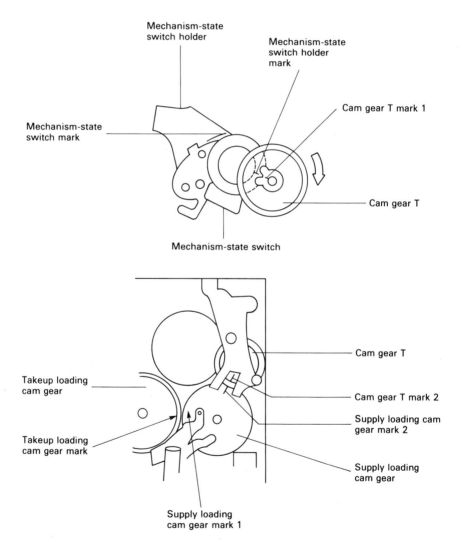

FIGURE 8-10 Mechanism-state (mode-sense) switch adjustment

Sec. 8-2 Electrical Adjustments

5. Play back a monoscope signal on an alignment tape.
6. Adjust the X-value adjustment point in either direction for maximum head envelope.
7. Tighten screws D and E, and apply lock paint to screw D.

8-1.15 Mechanism-State (Mode-Sense) Switch Adjustment

Figure 8-10 is the adjustment diagram.

The purpose of this adjustment is to align the state switches in relation to the loading mechanism position. If there is any misalignment, the state switch will send false loading mechanism position information to system-control microprocessor IC901.

1. Align the supply loading cam gear mark 1 with the mark located on the takeup loading cam gear.
2. Check that the markings on the mechanism-state switch and cam gear T are aligned. If the markings are not aligned, reinstall the mechanism-state switch. Note that this adjustment is generally not necessary unless parts have been replaced (or there has been tinkering).
3. Turn the cam gear T clockwise (direction of the arrow), and align the markings on the mechanism-state switch and the switch holder.
4. Check that the markings on the supply and takeup loading gears are aligned. If the markings are not aligned, reinstall the loading gears.
5. Install the mechanism-state switch holder into the chassis so that the markings on cam gear T and the supply loading gear are aligned.

8-2 ELECTRICAL ADJUSTMENTS

The following paragraphs describe complete adjustment procedures for the electrical sections of a typical camcorder. The test equipment and fixtures required for these adjustments are discussed in Sec. 3-2.

8-2.1 Test Equipment and Setup

Figure 3-1 shows a typical setup for camcorder electrical adjustments. The following steps should be accomplished before making any adjustments.

1. Connect the camcorder and test equipment as shown in Fig. 3-1.
2. Allow the camcorder and test equipment to warm up for at least 30 min before making the adjustments. Unless otherwise specified, the illumination on the chart should be 1400 lux.

3. Preset the camcorder controls as follows: playback/camera to camera, standby to on, autofocus to manual, auto iris to manual, white balance to center or off.
4. *Do not* set the automatic white-balance control to on or auto during the adjustment procedure. If the white-balance control is accidentally set to auto during the adjustment procedure, turn the power supply off, cap the lens, and set the white-balance switch to the center position, before turning the power supply on again.
5. Connect a 2.2k-ohm resistor between pin 25 of IC1104 and ground. The electrical position of pin 25 on IC1104 is shown in Fig. 4-21. This resistance disables the white-balance function.
6. Set the flutter control RT1330 to the center position. Figure 4-9 shows the electrical position of RT1330.

8-2.2 Burst Frequency Adjustment

The purpose of this adjustment is to set the burst or subcarrier oscillator at the correct frequency. Misadjustment of the burst or subcarrier oscillator can cause no-color, no-color-sync, and tint (wrong color) problems.

1. Connect a frequency counter to TP1115.
2. Adjust the burst or subcarrier oscillator control CT1101 for a reading of 3.579545 MHz ± 20 Hz on the frequency counter.

Figure 4-16 shows the electrical position of CT1101, while Fig. 4-26 shows TP1115.

8-2.3 Dynamic Focus Release

Figure 4-12 shows the electrical position of the controls and test points. Figure 8-11 shows the waveforms.

1. Cap the lens.
2. Connect an oscilloscope to TP1503.
3. Trigger the oscilloscope at the horizontal rate.
4. Adjust the dynamic focus H. saw control RT1505 and the dynamic focus H. para control RT1506 to make the waveform flat.
5. Trigger the oscilloscope at the vertical rate.
6. Adjust the dynamic focus V. saw control RT1507 and the dynamic focus V. para control RT1508 to make the waveform flat.

Figure 8-11 shows both good and bad waveforms for horizontal and vertical signals.

Sec. 8-2   Electrical Adjustments   249

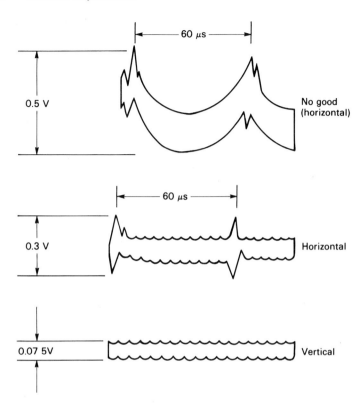

FIGURE 8-11 Dynamic focus release waveforms

8-2.4 R- and B-Signal Shading Release

Figure 4-23 shows the electrical position of the controls and test points. Figure 8-12 shows the waveforms.

1. Cap the lens.
2. Connect the scope channel 1 probe to TP1505.
3. Connect the scope channel 2 probe to TP1504.
4. Trigger the scope at the horizontal rate.
5. Adjust the red shading H. saw control RT1513 and the red shading H. para control RT1514 to make the waveform at TP1505 flat.
6. Adjust the blue shading H. saw control RT1509 and the blue shading H. para control RT1510 to make the waveform at TP1504 flat.
7. Trigger the scope at the vertical rate.
8. Adjust the red-shading V. saw control RT1515 and the red-shading V. para control RT1516 to make the waveform at TP1505 flat.

9. Adjust the blue-shading V. saw control RT1511 and the blue-shading V. para control RT1515 to make the waveform at TP1504 flat.

8-2.5 R- and B-Signal Tracking Release

Figure 4-23 shows the electrical position of the controls and test points.

1. Cap the lens.
2. Set controls RT1107 through RT1112 to their mechanical centers.

8-2.6 Target Voltage Adjustment

Figure 4-15 shows the electrical position of the controls and test points. The purpose of this adjustment is to set the voltage on the Saticon tube target

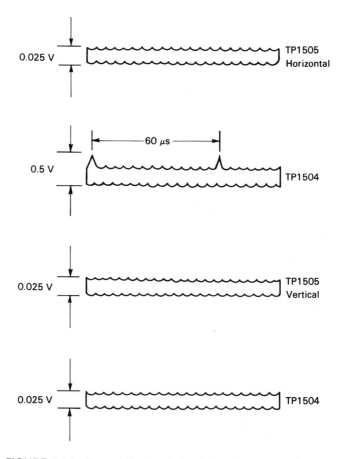

FIGURE 8-12 R- and B-signal shading release waveforms

Sec. 8-2 Electrical Adjustments   251

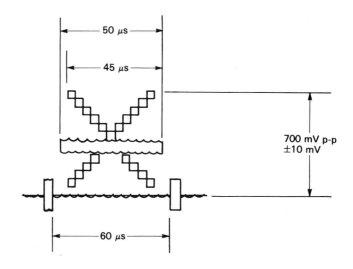

FIGURE 8-13 AIC level adjustment waveforms

at the correct value. A total loss of target voltage causes a no-picture condition. A low-target voltage can cause a weak picture.

1. Cap the lens.
2. Connect a DVM to TP1306.
3. Adjust the target voltage control RT1322 for 41 ± 1 V.

8-2.7 AIC Level Adjustment

Figures 4-19 and 4-26 show the electrical position of the controls. Figure 8-13 shows the waveforms. This adjustment sets the balance point of the automatic iris. Depending on the AIC setting (too high or too low) an incorrect adjustment can cause the image to appear dark or bright under normal lighting.

1. Aim the camera at the grayscale chart (Fig. 3-1).
2. Adjust the zoom ring to fill the screen with the grayscale chart.
3. Connect an oscilloscope to the video output at the A/V output connector.
4. Trigger the oscilloscope at the horizontal rate.
5. Turn the chroma gain control RT1120 (Fig. 4-26) fully counterclockwise (as seen from the component side).
6. Adjust the AIC control RT1102 (Fig. 4-19) for 700 ± 10 mV p-p (Fig. 8-13).

After the adjustment is complete, turn the chroma gain control RT1120 to the mechanical center.

8-2.8 AGC Maximum Gain Adjustment

Figure 4-18 shows the electrical position of the controls and test points. Figure 8-14 shows the waveforms.

1. Aim the camera at the grayscale chart.
2. Connect an oscilloscope to TP1101 through an 18K-ohm resistor.
3. Trigger the oscilloscope at the horizontal rate.
4. Adjust low-light gain control RT1103 so that the waveform is 600 ± 50 mV p-p.

8-2.9 Beam Current Adjustment

Figures 4-14, 4-15, and 4-18 show the electrical position of the controls and test points. Figure 8-15 shows the waveforms. This procedure adjusts the Saticon electron beam by varying the G1 (grid 1) voltage. When the voltage is too high, the shading increases. When the voltage is too low, the picture tends to wash out or not appear at all.

1. Connect TP1116 to ground.
2. Connect an oscilloscope to TP1101.
3. Trigger the oscilloscope at the horizontal rate.
4. Aim the camera at a high-intensity object. Make sure the iris is open, and adjust the zoom ring so that the light fills a small area in the center of the monitor TV screen.

Caution: Do not leave the camera aimed at high-intensity objects for long periods of time.

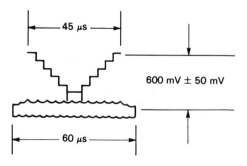

FIGURE 8-14 AGC maximum gain adjustment waveforms

Sec. 8-2 Electrical Adjustments   253

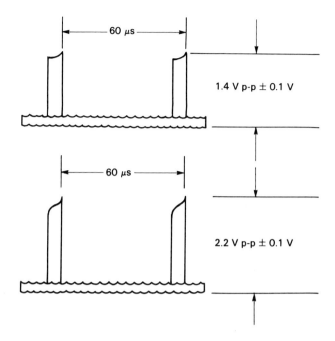

FIGURE 8-15 Beam current adjustment waveforms

5. Adjust beam control RT 1324 until the waveform is clipped at 1.4 ± 0.1 V p-p.
6. Disconnect TP1116 from ground.
7. Adjust ABO control RT1325 until the waveform is clipped at 2.2 ± 0.1 V p-p.

8-2.10 Streaking Adjustment

Figure 4-17 shows the electrical position of the controls. Figure 8-16 shows the grayscale chart display on the monitor TV. This adjustment matches the white, cyan, and yellow delay lines to minimize streaking in the horizontal direction.

1. Aim the camera at the grayscale chart and adjust the monitor TV color level control to maximum.
2. Adjust streaking control CT1001 to minimize color streaking in the black-to-gray and white-to-gray transitions.

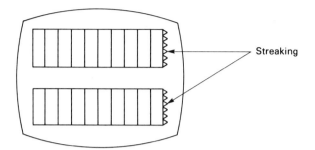

FIGURE 8-16 Grayscale chart display on monitor TV (with streaking)

8-2.11 Dark-Shading Adjustment

Figures 4-18 and 4-19 show the electrical positions of the controls and test points. Figure 8-17 shows the waveforms. This adjustment corrects shading of the bias light so that the whole picture becomes dark without color noise when the camera is aimed at a dark area.

1. Cap the lens.
2. Connect an oscilloscope to TP1103 through a 18k-ohm resistor.
3. Trigger the oscilloscope at the vertical rate.
4. Adjust the dark shading V. saw control RT1503 and the dark shading V. para control RT1504 so that the waveform is flat.
5. Trigger the oscilloscope at the horizontal rate.
6. Adjust the dark shading H. saw control RT1501 and the dark shading H. para control RT1502 so that the waveform is flat.

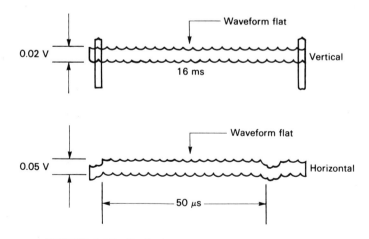

FIGURE 8-17 Dark-shading adjustment waveforms

Sec. 8-2   Electrical Adjustments   255

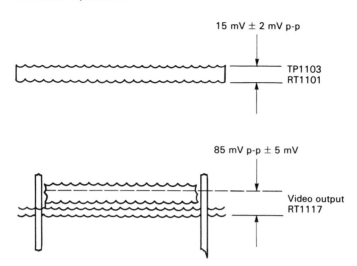

FIGURE 8-18   Dark-offset (black level) adjustment waveforms

8-2.12   Dark-Offset Adjustment

Figures 4-18, 4-19, and 4-20 show the electrical positions of the controls and test points. Figure 8-18 shows the waveforms. This adjustment sets the black level.

1. Cap the lens.
2. Connect an oscilloscope to TP1103.
3. Trigger the oscilloscope at the horizontal rate.
4. Adjust AGC offset control RT1101 for 15 ± 2 mV p-p from the blanking level to the center of the waveform.
5. Connect the oscilloscope to the video output (at the A/V output connector).
6. Adjust the Y setup control RT1117 for 85 ± 5 mV p-p from the blanking level to the center of the waveform.

8-2.13   Backfocus Adjustment

Figure 8-19 is the adjustment diagram. Figure 3-1 shows a typical backfocus chart. The purpose of this adjustment is to ensure proper focus tracking throughout the zoom range.

1. Set chart illumination between 50 and 100 lux (for this adjustment).
2. Aim the camera at the backfocus chart. Position the camera 10 ft (3 m) from the backfocus chart.

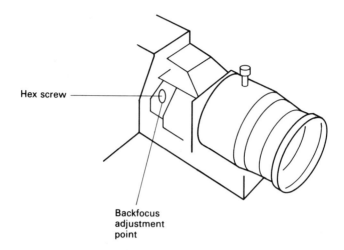

FIGURE 8-19 Backfocus adjustment

3. Set the lens to the 10 ft (3 m) position.
4. Loosen the hexagonal screw shown in Fig. 8-19.
5. Zoom in (close-up/telephoto) and adjust the lens focus for best focus on the monitor TV.
6. Zoom out (wide angle) and insert the backfocus adjustment driver into the backfocus adjustment point. (A special backfocus adjustment tool is required for this adjustment, on most camcorders.) Adjust the backfocus for best focus on the monitor TV.
7. Zoom in again, and adjust the lens for best focus.
8. Zoom out again, and adjust the backfocus for best focus.
9. Repeat the procedure until the focus tracks throughout the zoom range.

8-2.14 Chroma Carrier Level Adjustment

Figures 4-15 and 4-21 show the electrical positions of the controls and test points. Figure 8-20 shows the waveforms. The purpose of this procedure is to adjust the Saticon focus electrode voltage so the beam is focused on the photoconductive film and to adjust the carrier level of the chroma signal. Incorrect level setting can cause color and tint problems.

1. Aim the camera at the grayscale chart (Fig. 3-1).
2. Connect the oscilloscope to TP1108.
3. Trigger the oscilloscope at the horizontal rate.
4. Adjust focus control RT1323 so that the waveforms are maximum.

Sec. 8-2 Electrical Adjustments   257

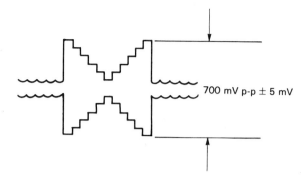

FIGURE 8-20   Chroma carrier level adjustment waveforms

5. Adjust carrier gain control RT1104 so that the waveform is 700 ± 50 mV p-p.

8-2.15 Vertical Size Adjustment

Figures 4-9, 4-10, and 4-21 show the electrical positions of the controls and test points. Figure 8-21 shows the waveforms. This adjustment sets the proper vertical scanning on the Saticon. Improper adjustment can cause distorted images or space at the top or bottom of the picture.

1. Mark the center of the monitor TV screen.
2. Connect TP1310 to a source of standby 5 V.
3. Adjust V-center control RT1321 so the vertical center mark on the Saticon tube is aligned with the mark on the monitor TV screen.
4. Disconnect TP1310 from the standby 5 V.
5. Aim the camera at the lightbox or white chart, and adjust the zoom ring until the lightbox fills the monitor TV screen.
6. Connect the oscilloscope to TP1108.
7. Trigger the oscilloscope at the vertical rate.
8. Adjust V-size control RT1320 for minimum beat during the vertical period.

Note that Fig. 8-21 shows both good and bad waveforms.

8-2.16 Horizontal Size and Linearity Adjustment

Figures 4-9 and 4-19 show the electrical positions of the controls and test points. Figure 8-22 shows the waveforms. The purpose of this adjust-

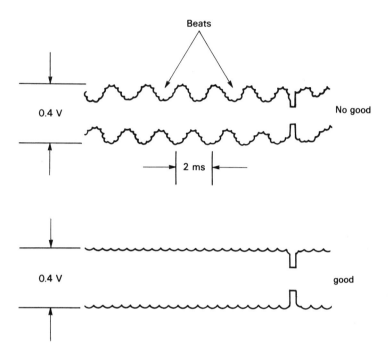

FIGURE 8-21 Vertical size adjustment waveforms

ment is to set the proper horizontal scanning on the Saticon. Improper adjustment can cause distorted images or a shift in color balance.

1. Aim the camera at the lightbox or white chart.
2. Connect the oscilloscope to TP1103.
3. Trigger the oscilloscope at the vertical rate.
4. Delay the waveform and observe the waveform center.
5. Adjust H. size control RT1317 so that the 4.3-MHz carrier in the center portion of the waveforms is minimum.
6. Adjust H. top linear control RT1318 for minimum carrier over all the signal.
7. Adjust H. linear control L1302 so the carrier at the beginning portion of the signal is minimum and flat.

Note that Fig. 8-22 shows both good and bad waveforms.

8-2.17 Horizontal and Vertical Center Adjustment

Figures 4-9 and 4-10 show the electrical positions of the controls and test points. This adjustment determines the horizontal and vertical deflection

Sec. 8-2 Electrical Adjustments

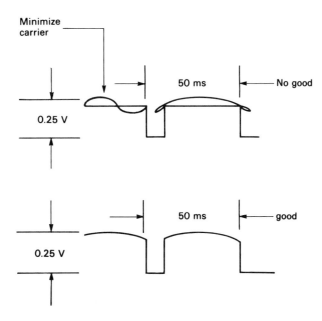

FIGURE 8-22 Horizontal size and linearity adjustment waveforms

position on the Saticon. If misadjusted, a space can appear on either side of the picture or at top or bottom.

1. Aim the camera at a focus chart.
2. Mark the center of the monitor TV screen.
3. Adjust the camera position so that a circle on the chart (or a center cross line) is aligned with the center of the monitor TV screen.
4. Zoom in and zoom out while adjusting H. center control RT1319 and V. center control RT1321, until the center mark on the monitor screen remains in the center of the circle or cross throughout the zoom range.

8-2.18 Vertical-Edge Correction Adjustment

Figure 4-22 shows the electrical positions of the controls and test points. Figure 8-23 shows the waveforms. This adjustment corrects the "colored-edge phenomenon" that occurs horizontally in the line following a dark-to-light or light-to-dark transition.

1. Aim the camera at the grayscale chart.
2. Connect the oscilloscope to TP1118.
3. Trigger the oscilloscope at the horizontal rate.

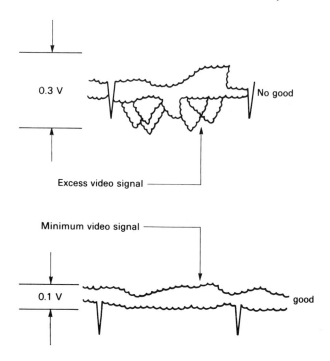

FIGURE 8-23 Vertical-edge correction adjustment waveforms

4. Adjust V. edge balance control RT1118 to minimize the video signal.

Note that Fig. 8-23 shows both good and bad waveforms.

8-2.19 Chroma Setup and Adjustment

Figure 4-25 shows the electrical positions of the controls and test points. Figure 8-24 shows the waveforms. This adjustment reduces color noise in dark areas.

1. Cap the lens.
2. Connect an oscilloscope channel 1 probe to TP1114 through an 18k-ohm resistor.
3. Connect the oscilloscope channel 2 probe to TP1113 through an 18k-ohm resistor.
4. Trigger the oscilloscope at the horizontal rate.
5. Adjust blue setup control RT1116 so that the chroma level at TP1113 matches the blanking level.

Sec. 8-2   Electrical Adjustments   261

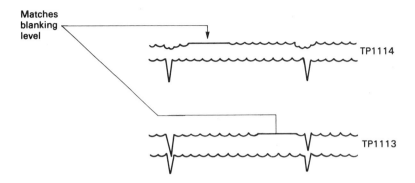

FIGURE 8-24 Chroma setup adjustment waveforms

6. Adjust the red setup control RT1115 so that the chroma level at TP1114 matches the blanking level.

Note that Fig. 8-24 shows both good and bad waveforms.

8-2.20 R- and B-Signal Separation Preset

Figures 4-21 and 4-25 show the electrical positions of the controls and test points. Figure 8-25 shows the waveforms.

1. Aim the camera at the lightbox or white chart.
2. Connect an oscilloscope channel 1 probe to TP1114.
3. Connect the oscilloscope channel 2 probe to TP1113.
4. Trigger the oscilloscope at the horizontal rate.
5. Adjust separation phase control RT1106 and separation gain control RT1105 for minimum flicker in the waveforms.

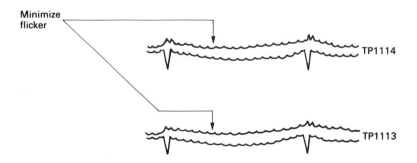

FIGURE 8-25 R- and B-signal separation preset waveforms

8-2.21 R- and B-Signal Gain Preset

Figures 4-24 and 4-25 show the electrical positions of the controls and test points. Figure 8-26 shows the waveforms.

1. Aim the camera at the lightbox or white chart.
2. Connect an oscilloscope channel 1 probe to TP1114.
3. Connect the oscilloscope channel 2 probe to TP1113.
4. Trigger the oscilloscope at the horizontal rate.
5. Adjust blue gain control RT1114 so the top of the chroma level at TP1114 matches the blanking level.
6. Adjust red gain control RT1113 so the top of the chroma level at TP1114 matches the blanking level.

8-2.22 Focus Preset Adjustment

Figures 4-15, 4-19, 4-21, and 4-25 show the electrical positions of the controls and test points. Figure 8-27 shows the waveforms.

1. Aim the camera at the grayscale chart.
2. Connect TP1105 to ground through a 47 μF/16 V electrolytic capacitor.
3. Connect an oscilloscope channel 1 probe to TP1114.
4. Connect the oscilloscope channel 2 probe to TP1113.
5. Trigger the oscilloscope at the horizontal rate.
6. Adjust focus control RT1323 so that the gray level of the waveform is straight and the first through sixth steps from the top are maximum.

If the waveform fluctuates regardless of correct synchronization during this adjustment, adjust RT1105 and RT1106 so that the fluctuations of the

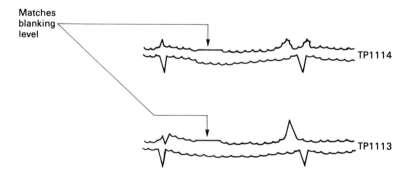

FIGURE 8-26 R- and B-signal gain preset waveforms

Sec. 8-2 Electrical Adjustments   263

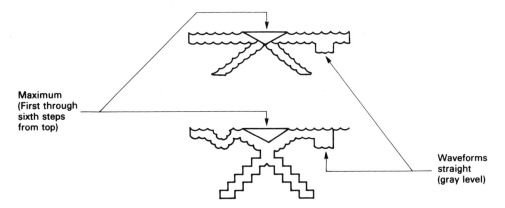

FIGURE 8-27 Focus preset adjustment waveforms

waveform are minimum. Then adjust RT1323 as described in step 6. Make certain to remove the capacitor from TP1105.

8-2.23 Dynamic Focus Preset

Figures 4-12 and 4-25 show the electrical positions of the controls and test points. Figure 8-28 shows the waveforms.

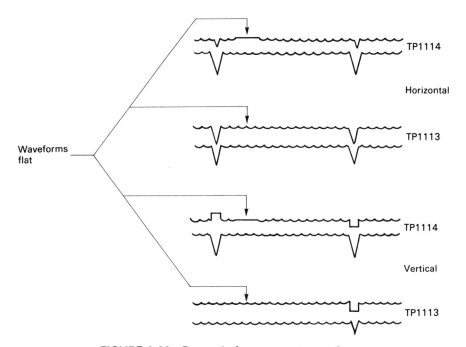

FIGURE 8-28 Dynamic focus preset waveforms

1. Aim the camera at the lightbox or white chart, and zoom in to fill the monitor TV screen.
2. Connect an oscilloscope channel 1 probe to TP1114.
3. Connect the oscilloscope channel 2 probe to TP1113.
4. Trigger the oscilloscope at the horizontal rate.
5. Adjust the dynamic focus H. para control RT1506 and the dynamic focus H. saw control RT1505 to make the waveform flat.
6. Trigger the oscilloscope at the vertical rate.
7. Adjust dynamic focus V. para control RT1508 and the dynamic focus V. saw control RT1507 to make the waveforms flat.

Repeat steps 4 through 7 for best results. There is usually some interaction between these adjustments.

8-2.24 R-Signal Shading Preset

Figures 4-23 and 4-25 show the electrical positions of the controls and test points. Figure 8-29 shows the waveforms.

1. Aim the camera at the lightbox or white chart, and zoom in to fill the monitor TV screen.
2. Connect an oscilloscope to TP1113.
3. Trigger the oscilloscope at the horizontal rate.
4. Adjust red-shading H. saw control RT1513 and red-shading H. para control RT1514 to make the waveform flat.
5. Trigger the oscilloscope at the vertical rate.
6. Adjust red-shading V. saw control RT1515 and red-shading V. para control RT1516 to make the waveform flat.

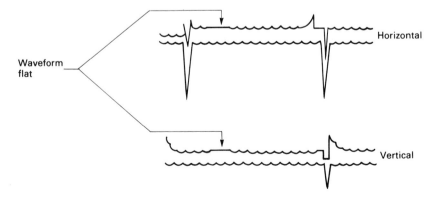

FIGURE 8-29 R-signal shading preset waveforms

Sec. 8-2   Electrical Adjustments   265

8-2.25 B-Signal Shading Preset

Figures 4-23 and 4-25 show the electrical positions of the controls and test points. Figure 8-29 shows the waveforms. This procedure is essentially the same as that for the R-signal shading preset (Sec. 8-2.24), except that the blue-shading controls RT1509 through RT1512 are used and the signal is monitored at TP1114.

8-2.26 Tracking Preset

Figures 4-23, 4-24, and 4-25 show the electrical positions of the controls and test points. Figure 8-30 shows the waveforms.

1. Aim the camera at the grayscale chart.
2. Connect an oscilloscope to TP1114.
3. Connect the oscilloscope to TP1113.
4. Trigger the oscilloscope at the horizontal rate.
5. Adjust blue gain control RT1114, blue tracking 1 control RT1110, blue tracking 2 control RT1111, and blue tracking 3 control RT1112 to make the waveforms flat and to match the blanking level at TP1114.
6. Adjust red gain control RT1113, red tracking 1 control RT1107, red tracking 2 control RT1108, and red tracking 3 control RT1109 to make the waveforms flat and to match the blanking level at TP1113.

8-2.27 Focus Adjustment

Figures 4-15, 4-19, and 4-25 show the electrical positions of the controls and test points. Figure 8-31 shows the waveforms.

1. Aim the camera at the grayscale chart.

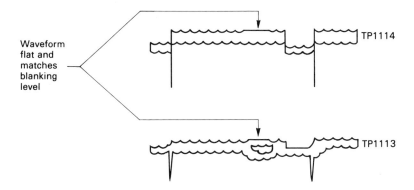

FIGURE 8-30 Tracking preset waveforms

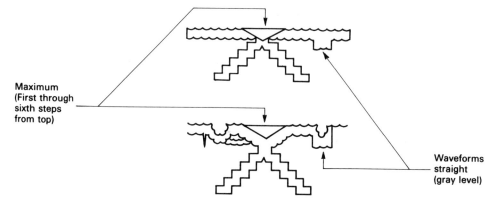

FIGURE 8-31 Focus adjustment waveforms

2. Connect TP1105 to ground through 47 μF/16 V electrolytic capacitor.
3. Connect an oscilloscope channel 1 to TP1114.
4. Connect the oscilloscope channel 2 to TP1113.
5. Trigger the oscilloscope at the horizontal rate.
6. Adjust focus control RT1323 so that the gray level of the waveform is straight and the first through sixth steps from the top are maximum. Make certain to remove the capacitor from TP1105.

8-2.28 Chroma Carrier Level Confirmation

Figure 4-21 shows the electrical positions of the controls and test points. Figure 8-20 shows the waveforms.

1. Aim the camera at the grayscale chart.
2. Connect an oscilloscope to TP1108.
3. Trigger the oscilloscope at the horizontal rate.
4. Confirm that the waveforms is 700 ± 40 mV p-p. If the level is not within the tolerance, readjust chroma carrier level with RT1104.

8-2.29 Dynamic Focus Adjustment

Figures 4-12 and 4-25 show the electrical positions of the controls and test points. Figure 8-32 shows the waveforms.

1. Aim the camera at the lightbox or white chart.
2. Connect an oscilloscope channel 1 to TP1114.
3. Connect the oscilloscope channel 2 to TP1113.

Sec. 8-2   Electrical Adjustments   267

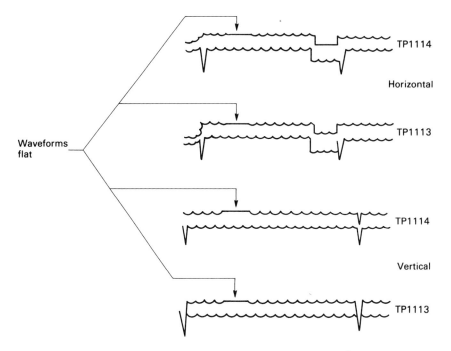

FIGURE 8-32   Dynamic focus adjustment waveforms

4. Trigger the oscilloscope at the horizontal rate.
5. Adjust dynamic focus H. para control RT1506 and dynamic focus H. saw control RT1505 to make the waveforms flat.
6. Trigger the oscilloscope at the vertical rate.
7. Adjust dynamic focus V. para control RT1508 and dynamic focus V. saw control RT1507 to make the waveform flat.

Repeat steps 4 through 7 for best results. There is usually some interaction between these adjustments.

8-2.30   R-Signal Shading Adjustment

This procedure is essentially the same as for the R-signal shading preset (Sec. 8-2.24), except that this is a final adjustment.

8-2.31   B-Signal Shading Adjustment

This procedure is essentially the same as for the B-signal shading preset (Sec. 8-2.25), except that this is a final adjustment.

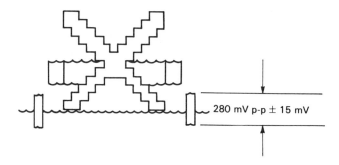

FIGURE 8-33 Burst level adjustment waveforms

8-2.32 Tracking Adjustment

This procedure is essentially the same as for the tracking preset (Sec. 8-2.26), except that this is a final adjustment. The procedure ensures that the red and blue signals track properly with the Y-signal at all light levels.

8-2.33 Burst Level Adjustment

Figure 4-26 shows the electrical positions of the controls and test points. Figure 8-33 shows the waveforms. This adjustment determines the level of color saturation.

1. Aim the camera at the grayscale chart.
2. Connect an oscilloscope to the video output at the A/V output connector. Terminate in 75 ohms.
3. Trigger the oscilloscope at the horizontal rate.
4. Adjust burst level control RT1121 for 280 ± 15 mV p-p.

8-2.34 Chroma Gain and Phase Adjustment

Figure 4-26 shows the electrical positions of the controls and test points. Figure 8-34 shows the waveforms.

1. Aim the camera at the grayscale chart, and optimize the white balance using the manual white-balance control. Note that up to this point in the procedure, the white balance has been off or in manual, but not in auto, as discussed in Sec. 8-2.1.
2. Aim the camera at the color chart.
3. Connect the video output from the A/V output connector to a vectorscope. Terminate in 75 ohms.

Sec. 8-2   Electrical Adjustments 269

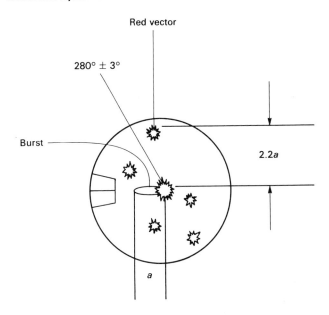

FIGURE 8-34 Chroma gain and phase adjustment vectorscope display

4. Adjust chroma gain control RT1120 so that the red vector is equal to 220% ± 10% when compared to the burst level.
5. Adjust burst phase control CT1102 so that the red vector is 280° ± 3° from the burst.

8-2.35 Flutter Cancel Adjustment

Figure 4-9 shows the electrical positions of the controls and test points.

1. Aim the camera at a white object outdoors.
2. If flutter appears in the EVF or monitor TV screen, adjust flutter control RT1330 to minimize the flutter.
3. After completing the flutter adjustment, recheck the horizontal size and linearity adjustments as described in Sec. 8-2.16.

8-2.36 Low-Light Chroma Gain Adjustment

Figure 4-26 shows the electrical positions of the controls and test points. Note that this adjustment is somewhat similar to the chroma gain and phase adjustment (Sec. 8-2.34) except that the adjustment is made under low-light conditions.

1. Reduce the chart illumination from 1400 lux to 25 lux.
2. Aim the camera at the color chart, and position the chart so the red portion of the chart is located in the center of the monitor TV screen.
3. Position the chart two ft from the camera lens.
4. Move the zoom in the telephoto direction until the red portion of the chart occupies one-third of the screen.
5. Cover the white portion of the color chart with black paper.
6. Connect the video output from the A/V output connector to a vectorscope. Terminate in 75 ohms.
7. Adjust chroma killer level control RT1123 so that the red vector is equal to 35–50% when compared to the burst level.

8-2.37 White-Balance Confirmation

1. Move the white-balance switch from manual to auto, and check that the picture displayed on the monitor TV does not change substantially.
2. If a considerable change is noticed when you switch from manual to auto white balance, recheck chroma setup (Sec. 8-2.10), R- and B-signal separation (Sec. 8-2.20), R and B gain (Sec. 8-2.21), and tracking (Sec. 8-2.26) adjustments.

Note that the white-balance control should be set for optimum in manual *before* switching to auto (as discussed in Sec. 8-2.34). If the white-balance control is set to an extreme position, there should be a substantial change when moving from manual to auto.

Also note that this white-balance check completes camera adjustments. Remove the resistor connected between pin 25 of IC1104 and ground (discussed in Sec. 8-2.1).

8-2.38 EVF Horizontal Hold Adjustment

Figure 4-32 shows the electrical positions of the controls and test points. This procedure sets the horizontal sync frequency during VCR playback.

1. Connect a frequency counter to TP1801.
2. Set horizontal hold control RT1801 to the mechanical center.
3. Connect a 47 μF/16 V electrolytic capacitor between the video input line and ground.
4. Adjust horizontal hold control RT1801 for 15.75 ± 0.1 kHz.
5. Disconnect the capacitor from the video line.

Sec. 8-2 Electrical Adjustments 271

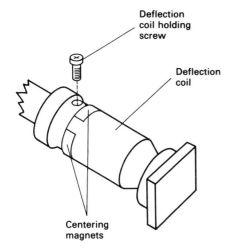

FIGURE 8-35 Relationship of deflection coil, holding screw, and centering magnets on EVF

8-2.39 EVF Picture Tilt Adjustment

Figure 8-35 shows the relationship of the deflection coil, holding screw, and centering magnets. This adjustment sets the picture tilt on the EVF screen.

1. *Make certain* that the camera is aligned with the object being viewed. The cross chart (Fig. 3-1), or some similar chart with both horizontal and vertical lines, is the best object for this adjustment. Both the camera and chart should be checked with a level (if practical).
2. Loosen the deflection coil holding screw.
3. Observe the EVF screen and turn the deflection coil until the image is straight.
4. Tighten the deflection coil holding screw.

8-2.40 EVF Centering Magnets Adjustment

Figure 8-35 shows the relationship of the deflection coil, holding screw, and centering magnets. This adjustment sets the center of the image picked up by the camera to the center of the EVF screen.

1. Aim the camera at a chart as described in Sec. 8-2.39.
2. Adjust the centering magnets until the center of the picture (cross chart) is in the center of the EVF screen.

8-2.41 EVF Vertical Size Adjustment

Figure 4-32 shows the electrical position of the controls and test points. This adjustment determines the vertical deflection size.

1. Aim the camera at a resolution chart or a cross chart with circle (Fig. 3-1).
2. Adjust vertical size control RT1802 so that the circle in the chart is round on the EVF screen.

8-2.42 EVF Focus Adjustment

Figure 4-32 shows the electrical position of the controls and test points. This adjustment sets the focus for the EVF.

1. Aim the camera at a resolution chart or a cross chart with circle (or the Siemens star chart shown in Fig. 3-1).
2. Adjust focus control RT1803 for best focus on the EVF screen.

8-2.43 EVF Contrast and Brightness Adjustment

Figure 4-32 shows the electrical position of the controls and test points. This adjustment sets contrast and brightness for the EVF.

1. Aim the camera at the grayscale chart.
2. Adjust brightness control RT1805 and the contrast control RT1804 so that the grayscale graduations seen on the EVF screen match those seen on a monitor TV screen.

8-2.44 Autofocus Sync Pulse-Width Confirmation

Figure 4-30 shows the electrical position of the controls and test points. Figure 8-36 shows the waveforms.

1. Connect an oscilloscope to TL4.
2. Trigger the oscilloscope at the vertical rate.
3. Check that the sync pulse period is 118 ± 15 μs.

If the sync pulse period is not within the specified tolerance, short or open SL-A and SL-B until the proper value is obtained.

Sec. 8-2 Electrical Adjustments   273

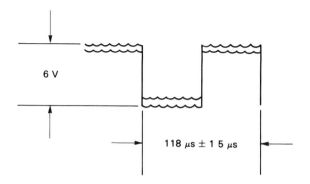

FIGURE 8-36 Autofocus sync-pulse-width confirmation waveforms

8-2.45 Autofocus Infrared Output Adjustment

Figure 4-30 shows the electrical position of the controls and test points. Figure 8-37 shows the waveforms.

1. Aim the camera at a black object.
2. Set the black object approximately 2 m from the camera.
3. Connect an oscilloscope to TL5 and TL6.
4. Check that the level is 3 ± 0.12 V.

If the level is incorrect, short or open SL-A and SL-B until the proper value is obtained.

8-2.46 Autofocus Offset Adjustment

Figure 4-30 shows the electrical position of the controls and test points. Figure 8-38 shows the waveforms.

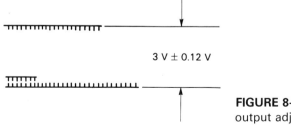

FIGURE 8-37 Autofocus infrared output adjustment waveform

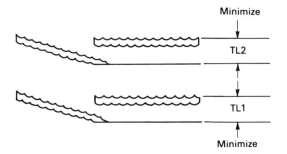

FIGURE 8-38 Autofocus offset adjustment waveforms

1. Cover the infrared receiver or sensor with thick black paper.
2. Turn the focus ring to the wide-angle position.
3. Connect an oscilloscope channel 1 to TL2.
4. Connect the oscilloscope channel 2 to TL1.
5. Trigger the oscilloscope at the vertical rate.
6. Adjust A channel offset control RT3 to minimize the level of the waveform at TL2.
7. Adjust B channel offset control RT2 to minimize the level of the waveform at TL1.

8-2.47 *Autofocus Gain Adjustment*

Figure 4-30 shows the electrical position of the controls and test points. Figure 8-39 shows the waveforms.

1. Aim the camera at a white object, and set the zoom to wide angle.
2. Position the white object 50 cm from the infrared receiver or sensor.
3. Cover the infrared receiver with a thin sheet of transparent paper (tracing paper).

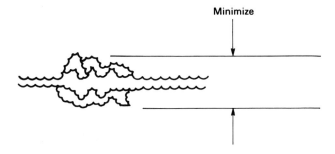

FIGURE 8-39 Autofocus gain adjustment waveforms

Sec. 8-2   Electrical Adjustments   275

4. Connect an oscilloscope channel 1 to TL1.
5. Connect the oscilloscope channel 2 to TL2. Set channel 2 to the invert mode.
6. Set the oscilloscope to the add mode.
7. Adjust gain-control RT1 so that the difference or level of the waveform is minimized.

8-2.48 Autofocus Motor-Speed Adjustment

Figure 4-31 shows the electrical position of the controls and test points.

1. Aim the camera at a white object.
2. Position the white object 1 m from the infrared receiver or sensor.
3. Set the zoom to telephoto.
4. Connect a DVM across both terminals of the motor.
5. Adjust motor speed control RT5 so that the voltage is 4.3 ± 0.1 V across the motor.

8-2.49 Autofocus Sensor Vertical Position Adjustment

Figure 8-40 shows the location of the controls and test points.

1. Aim the camera at an autofocus chart. Use the recommended autofocus chart or a Siemens star chart as shown in Fig. 3-1.
2. Place the chart 3 m from the camera.

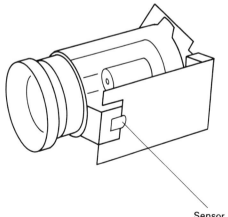

Sensor adjustment screw

FIGURE 8-40 Autofocus sensor vertical position adjustment controls

3. Adjust the sensor adjustment screw for optimum focus of the picture (as seen on the monitor TV). Note that the monitor TV must be in perfect focus before making this adjustment. Also, if you use the EVF to adjust camera focus, make certain that the EVF is properly focused first.

8-2.50 VCR 5-V Adjustment

Figure 7-5 shows the electrical position of the controls and test points. This adjustment sets the output voltage of the 5-V switching regulator IC906.

1. Apply 12 V to the external battery jack J950.
2. Place the camcorder in RECORD mode.
3. Connect a DVM to monitor the standby 5-V line.
4. Adjust 5-V adjust control RT951 for 5.3 ± 0.05 V.

8-2.51 VCR Overdischarge Level Adjustment

Figure 7-8 shows the electrical position of the controls and test points. This adjustment determines the overdischarge detection level. When the battery voltage drops to a specified level, the VCR automatically shuts off. When the overdischarge detection level is set below 11.5 V, the battery service life deteriorates. On the other hand, if the voltage is adjusted higher than 11.5 V, the usable time per charge is shortened.

1. Turn ODC control RT901 fully counterclockwise (viewed from the solder side).
2. Connect an external power supply, using the external battery jack J950.
3. Place the camcorder in RECORD mode.
4. Set the external supply to 11.6 ± 0.01 V.
5. Adjust ODC control RT901 clockwise until two of the battery display indicators turn off. In the circuit of Fig. 7-8, adjust RT901 until D1601 and D1602 just turn off.

8-2.52 VCR Cylinder Motor Speed Adjustment

Figures 5-12 and 5-15 show the electrical positions of the controls and test points. Figure 8-41 shows the waveforms. This adjustment locks the cylinder servo to the system signal and maintains a constant cylinder speed of 2700 rpm. When not adjusted properly, horizontal instability and/or picture noise increases.

1. Connect the oscilloscope to TP603.

Sec. 8-2   Electrical Adjustments   277

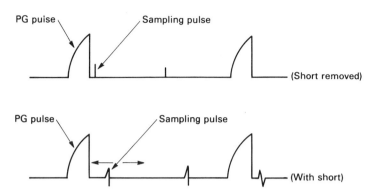

FIGURE 8-41   VCR cylinder motor-speed adjustment waveforms

2. Load the camcorder with a blank tape, and place the camcorder in the RECORD mode.
3. Connect free-run test point TP601 to standby 5 V. This locks the pulse-width modulator in IC601 to a fixed width and disables the cylinder phase-control circuit.
4. Adjust cylinder motor-speed control RT604 so that the sampling pulse stands still or moves slowly on the PG pulse.
5. Disconnect TP601 from the standby 5 V. Check that the sampling pulse is locked in on the trailing edge of the PG pulse.

8-2.53 VCR Capstan Motor-Speed Adjustment

Figures 5-15, 5-20, and 5-21 shows the electrical position of the controls and test points. Figure 8-42 shows the waveforms. This adjustment locks the capstan servo to the system signals and maintains a constant tape speed. When not adjusted properly, the vertical stability is degraded and/or picture noise increases.

1. Apply an NTSC colorbar signal to the EVF jack using a video/audio input adapter.
2. Connect the oscilloscope to TP604.
3. Load the camcorder with a blank tape, and place the camcorder in the RECORD mode.
4. Connect TP601 to standby 5 V.
5. Adjust capstan motor-speed control RT605 so that the sampling pulse stands still or moves slowly on the PG pulse.
6. Disconnect TP601 from the standby 5 V. Check that the sampling pulse is locked in on the trailing edge of the PG pulse.

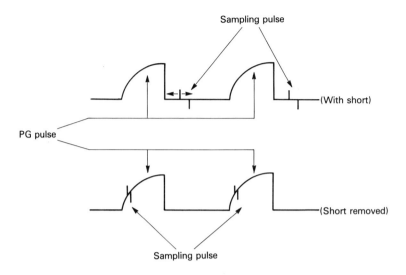

FIGURE 8-42 VCR capstan motor-speed adjustment waveforms

8-2.54 VCR 30-Hz Reference Frequency

Figure 5-21 shows the electrical positions of the controls and test points. This reference frequency is used to correct for fluctuations caused by unevenness among the belt, pulley, capstan motor, and capstan shaft.

1. Connect a frequency counter to the audio output at the A/V output connector.
2. Connect TP602 to standby 5 V.
3. Load the camcorder with an alignment tape, and play back the 3-kHz audio signal.
4. Check that the audio frequency is 3 kHz ± 15 Hz.
5. If the audio frequency is higher than 3015 Hz, remove resistor R613, and check that R612 is connected between pin 10 of IC601 and ground. (In some cases, the camcorders of Fig. 5-21 are shipped without R612 and/or R613. In such cases, add R612 if the audio frequency is high.)
6. If the audio frequency is lower than 2985 Hz, remove R612 and check that R613 is connected between pin 10 of IC601 and standby 5 V. If R613 is missing, and the audio frequency is low, add R613.

Note that both R612 and R613 are 10k-ohm in the camcorder of Fig. 5-21.

Sec. 8-2 Electrical Adjustments   **279**

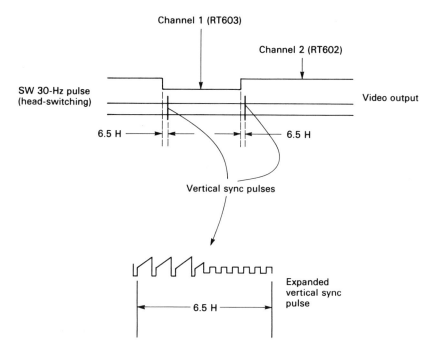

FIGURE 8-43   VCR PG shifter adjustment waveforms

8-2.55 VCR PG Shifter Adjustment

Figures 5-15 and 6-6 show the electrical positions of the controls and test points. Figure 8-43 shows the waveforms. This adjustment determines the head-switching point during playback. Misadjustment of the PG shifters may cause head-switching noise and/or vertical jitter in the picture.

1. Connect an oscilloscope channel 1 (1 V/div, 50 μs/div) to the video output using the A/V output connector.
2. Connect the oscilloscope channel 2 (1 V/div) to TP205 (30-Hz SW).
3. Load the camcorder with a colorbar alignment tape, and play back the colorbar portion of the tape.
4. Set the oscilloscope to ($-$) slope, and adjust channel 1 PG shifter control RT603 so that the trailing edge of the SW30 pulse is placed 6.5 ± 0.5 H (horizontal) lines before the start of the vertical sync pulse.
5. Set the oscilloscope to ($+$) slope, and adjust channel 2 PG shifter control RT602 so that the trailing edge of the SW30 pulse is placed 6.5 ± 0.5 H (horizontal) lines before the start of the vertical sync pulse.

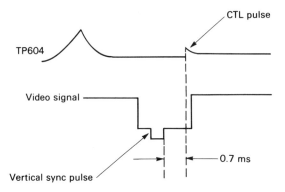

FIGURE 8-44 VCR tracking preset adjustment waveforms

8-2.56 VCR Tracking Preset Adjustment

Figure 5-21 shows the electrical positions of the controls and test points. Figure 8-44 shows the waveforms.

This adjustment ensures proper tracking of a tape recorded and played back on the same camcorder. Incorrect adjustment may cause noise in the picture.

1. Connect an oscilloscope channel 1 (1 V/div, 5 ms/div) to TP604. Trigger the oscilloscope on channel 1.
2. Connect the oscilloscope channel 2 (1 V/div) to the video output using the A/V output connector.
3. Apply an NTSC colorbar signal to the EVF jack using the video/audio input adapter.
4. Set the user tracking control to the detent or center position.
5. Load the camcorder with a blank tape, and record about 2 min of the colorbar signals.
6. Play back the recorded signals, and adjust tracking preset control RT601 so that the positive peak of the CTL pulse and the vertical sync pulse of the video signal are aligned as shown in Fig. 8-44.

8-2.57 VCR Record Chroma Level Adjustment

Figures 6-2 and 6-6 show the electrical positions of the controls and test points. Figure 8-45 shows the waveforms. This adjustment regulates the recorded color level. If the level is too high, diamond-shaped beats may be seen in the picture. If the level is too low, the color may be degraded.

1. Connect an oscilloscope (10 mv/div, 20 μs/div) to TP201.

Sec. 8-2 Electrical Adjustments 281

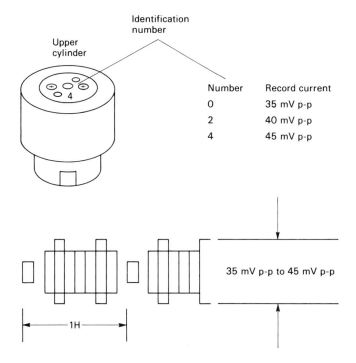

FIGURE 8-45 VCR record chroma level adjustment waveforms

2. Apply an NTSC colorbar signal (1 V p-p) to the EVF jack using the video/audio input adapter.
3. Check the upper cylinder for an identification number (0–4) stamped on the top (Fig. 8-45).
4. Turn the record luma level control RT201 fully clockwise (viewed from the component side).
5. Load the camcorder with a blank tape, and place the camcorder in the record mode.
6. Adjust record chroma level control RT202 for the value indicated on the chart in Fig. 8-45. For example, if the identification number stamped on the cylinder is 4, adjust RT202 for a record current of 45 mV p-p.
7. Perform the record luma level adjustment (Sec. 8-2.58).

8-2.58 VCR Record Luma Level Adjustment

Figures 6-2 and 6-6 show the electrical positions of the controls and test points. Figure 8-46 shows the waveforms. This adjustment sets the level of the record luma current at an optimum point. If the level is too high, video may overload. If the level is too low, the signal-to-noise ratio deteriorates.

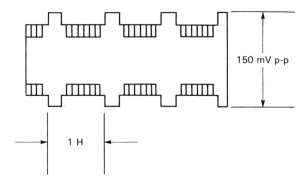

FIGURE 8-46 VCR record luma level adjustment waveforms

1. Perform the record chroma level adjustment described in Sec. 8-2.57 before making this adjustment.
2. Connect an oscilloscope (50 mv/div, 20 µs/div) to TP201.
3. Apply an NTSC colorbar signal (1 V p-p) to the EVF jack using the video/audio input adapter.
4. Load the camcorder with a blank tape, and place the camcorder in the RECORD mode.
5. Adjust record luma level control RT201 for a level of 150 mV p-p as shown in Fig. 8-46.

8-2.59 VCR Audio Playback Gain Adjustment

Figure 6-10 shows the electrical positions of the controls and test points. This adjustment sets the level of playback gain.

1. Connect a millivoltmeter (or oscilloscope) to the audio output using the A/V output connector.
2. Load the camcorder with an alignment tape, and play back a 1-kHz portion of the tape.
3. Adjust audio playback gain control RT401 for 90 mV rms or 255 mV p-p.

8-2.60 VCR Audio Bias Level Adjustment

Figure 6-9 shows the electrical positions of the controls and test points. This adjustment optimizes the audio record bias. If the bias is too low, high frequencies are increased, resulting in distortion. If the bias is too high, high frequencies are attenuated.

1. Connect a millivoltmeter between TP401 and TP402.

Sec. 8-2 Electrical Adjustments

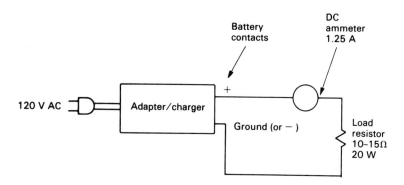

FIGURE 8-47 Adapter/charger current adjustment diagram

2. Load the camcorder with a blank tape, and place the camcorder in the RECORD mode (but with no audio signal applied).
3. Adjust audio bias level control RT402 for 2.6 mV rms ± 0.1 mV.

8-2.61 Adapter/Charger Current Adjustment

Figure 7-2 shows the electrical positions of the controls and test points. Figure 8-47 shows the test diagram. This adjustment sets the level of the adapter/charger output current.

1. Check that the a-c line voltage is 120 V.
2. Turn current adjust control R61 fully counterclockwise (from the component side).
3. Connect a load resistor (10 to 15 ohm, 20 W) between the positive (+) terminal (battery contacts) of the adapter/charger and ground, with a d-c ammeter in series with the load resistor.
4. Press charge-sense (or battery-sense) switch SW2, and adjust R61 until the current meter reads 1.25 A.

9
TROUBLESHOOTING and SERVICE NOTES

This chapter describes a series of troubleshooting procedures for a cross section of VHS camcorders. As discussed in the Preface, it is not practical to provide a specific troubleshooting procedure for every camcorder. Instead, we described a universal troubleshooting approach, using specific examples of camcorders. These examples just happen to be some of those camcorders discussed in Chapters 4 through 7. In this way, you can relate the theory (Chapters 4 through 7) to the troubleshooting procedures in this chapter; then you can relate both to the specific camcorder you are servicing.

9-1 THE BASIC TROUBLESHOOTING APPROACH

It is assumed that you are already familiar with the basics of electronic troubleshooting, including solid-state troubleshooting. If not, and you plan to service camcorders, you are in terrible trouble. Your attention is directed to the author's *Handbook of Advanced Troubleshooting*.

In the case of a camcorder, there are seven basic functions required for troubleshooting and repair.

First, you must study the camcorder using service literature, user instructions, schematic diagrams, and so on, to find out how each circuit works when operating normally. In this way you know in detail how a given camcorder should work. If you do not take the time to learn what is normal, you cannot tell what is abnormal. For example, some camcorders simply

Sec. 9-1   The Basic Troubleshooting Approach   285

take and play back better pictures than other camcorders, even when recording the same scene. You can waste hours of precious time (money) trying to make the inferior camcorder perform like a quality instrument if you do not know what is normal operation.

Second, you must know the function of, and how to manipulate, all camcorder controls and adjustments. (It is also assumed that you know how to operate the controls of the TV set used to monitor the camcorder playback. An improperly adjusted monitor TV can make a perfectly good camcorder appear bad.) In any case, it is difficult, if not impossible, to check out a camcorder or TV without knowing how to set the controls. Also, as camcorders and TVs age, readjustment and alignment of critical circuits are often required.

Third, you must know how to interpret service literature and how to use test equipment. Along with good test equipment that you know how to use, well-written service literature is your best friend. In general, camcorder service literature is excellent as far as procedures (operation, adjustment, disassembly, reassembly) and illustrations (drawings and photographs) are concerned. Unfortunately, camcorder literature is often weak when it comes to descriptions of how circuits operate, why circuits are needed, and so on (the theory of operation). The "how it works" portion of much camcorder literature is somewhat skimpy or simply omitted. This is especially true of late-model camcorder literature. Particularly annoying is where the literature says "Model X uses the same circuits as Model Y, except as follows," and then goes on to explain only a few Model X circuits. They assume that everyone knows how a camcorder works and/or that you have the literature for Model Y. (Or is it a scheme to sell more service manuals?!)

Fourth, you must be able to apply a systematic, logical procedure to locate troubles. Of course, a logical procedure for one type of camcorder is quite illogical for another. For example, it is quite illogical to check operation of the manual iris control on a camcorder not so equipped. However, it is quite logical to check operation of the automatic iris-control circuit on any camcorder. We discuss logical troubleshooting procedures for the various circuits of a camcorder at the end of the circuit descriptions in Chapters 4 through 7.

Fifth, you must be able to analyze logically the information of an improperly operating camcorder. The information to be analyzed may be in the form of performance, such as the appearance of the picture on the EVF or a known-good monitor TV, or may be indications taken from test equipment, such as waveforms monitored with an oscilloscope. Either way, it is your analysis of the information that makes for logical, efficient troubleshooting.

One problem in analyzing a camcorder during service is that camcorders must be played back through a TV for a complete check (not just through the EVF). If the TV is defective or improperly adjusted, the camcorder may appear to have troubles. For this reason, you must have at least one TV of known quality for evaluation of trouble symptoms. All camcorders passing through the shop can be compared against the same standard. The ultimate camcorder monitor is an industrial receiver/monitor. However, the most practical approach for the average shop is to use a monitor-type TV. This makes it possible to feed audio/video signals from the camcorder directly to the display circuits of the TV, bypassing the RF/IF/SIF circuits.

The obvious first move in analyzing a camcorder, once you have determined that there is a problem in the camcorder and not with the TV, is to play back an alignment tape or program tape of known quality. This effectively splits the camcorder circuits in half. (If you are a regular reader of the author's books, you know that the half-split is a standard technique in electronic troubleshooting. If you are not a regular reader, you should be!)

If playback is not satisfactory on either the EVF or monitor TV, the VCR section circuits are suspect. Of course, if the playback is good on the TV, but not on the EVF, you have localized the problem to the EVF circuits.

If playback is satisfactory on both the EVF and monitor TV (with a known-good tape), it is reasonable to assume that the playback circuits in the VCR section of the camcorder, and the EVF circuits, are good. In that case, the problem is likely in the camera circuits of the camcorder.

The next step is to insert a blank tape, record a scene, and play back the scene on the EVF and monitor TV. In the shop, it is best to record a scene with stable lighting and easy-to-compare colors, such as a color chart (Fig. 3-1) with fixed lighting.

If the camcorder is capable of playing back an alignment tape (or known-good tape), but not a scene taken by the same camcorder, the problem is likely in the camera circuits, but it could be in the record circuits of the VCR section.

As a final rough check, compare the scene shown on the EVF during record with the display during playback. If the scenes are essentially the same, it is reasonable to assume that the VCR section record circuits are good.

Sixth, you must be able to perform complete checkout procedures on a camcorder that has been repaired. Such checkout may be only a simple operation, such as selecting each operating mode in turn. At the other extreme, the checkout can involve complete readjustment of the camcorder, both electrical and mechanical. In any event, some checkout is required after any troubleshooting.

One practical reason for the checkout is that there may be more than one trouble. For example, an aging part may cause high current to flow through a resistor, resulting in burnout of the resistor. Logical troubleshooting may lead you quickly to the burned-out resistor. Replacement of the resistor can restore operation. However, only a thorough checkout can reveal the original high-current condition that caused the burnout.

Another reason for after-service checkout is that the repair may have produced a condition that requires readjustment. A classic example of this occurs where replacement of the video heads almost always requires readjustment of both electrical and mechanical components on the camcorder.

Seventh, you must be able to use proper tools to repair the trouble. Camcorder service requires all the common handtools and test equipment found in TV/VCR service, plus many special tools, jigs, fixtures, and charts that are unique to video cameras.

In summary, before troubleshooting any camcorder, ask yourself these questions: Have I studied all available service literature to find out how the camcorder works (including any special features such as remote control, RF adapters, character generators, etc.)? Can I operate the camcorder controls properly? Do I really understand the service literature, and can I use all required test equipment and tools properly? Using the service literature and/or previous experience on similar camcorders, can I plan out a logical troubleshooting procedure? Can I analyze logically the results of operating check as well as checkout procedures involving test equipment? Using the service literature and/or experience, can I perform complete checkout procedures on the camcorder, including realignment, adjustment, and so on, if necessary? Once I have found the trouble, can I use common handtools to make the repairs, such as handling electrostatically sensitive or ES devices, flat-back IC removal/replacement, leadless or chip component removal/replacement, and circuit board foil repair)? If the answer is no to any of these questions, you simply are not ready to start troubleshooting any camcorder. Start studying!

9-2 OPERATIONAL CHECKLIST

Before we get into detailed troubleshooting, let us review some simple, obvious steps to be performed first. The following checklist describes common symptoms and possible causes for some basic camcorder troubles. Make these checks before you tear into the camcorder with soldering tool and hacksaw.

Camcorder cannot be turned on. No power. First, make sure the power button is on, and that the standby button is not on. If you are using a battery, make sure the battery is fully charged. If you are using the adapter/charger, make sure the adapter/charger is properly plugged in. Check the dew indicator. The camcorder should not operate if the dew indicator is on. Make sure that all necessary cables (if any) are connected correctly and firmly. Finally, if power seems to be on, but you get no picture on the EVF, make sure the lens cap is off!

Cassette cannot be inserted. Check that the power button is on. Insert the cassette with the window side facing out, and the safety tab facing up.

Cassette cannot be ejected. Check that the power button is on. Then press the eject button to open the cassette compartment.

Camcorder cannot be operated in any mode. Check the dew indicator. The camcorder should not operate if the dew indicator is on.

Camcorder cannot be operated in record. Make sure that the safety tab on the back of the cassette is in place. Make sure the battery indicator in the EVF is not flashing. Make sure the dew indicator in the EVF is not flashing.

Focus is not sharp. Make sure the lens is properly focused. Set the focus switch to auto. Try focusing the lens manually. Also make sure the lens surface is not dirty or dusty. Note that if the focus is good on a monitor TV, but not on the EVF, you have problems in the EVF focus circuits (which are separate from the camera focus circuits).

Color balance is not proper. Check if the condition can be corrected by adjustment of white balance (either automatic or manual).

Color picture is not satisfactory. Make certain that the monitor-TV color circuits are properly adjusted. If the camcorder output is being fed to the monitor TV through an RF adapter (to the TV antenna terminals instead of the video-in jack), make certain that the TV fine-tuning is properly adjusted.

Playback picture is noisy or contains streaks. Try correcting the condition with the tracking control.

Playback picture shows blurred action or indicates motion. If this occurs only when the subject is moving rapidly (or when the camcorder is being panned rapidly), the condition may be normal (even for camcorders with MOS pickups, which are supposed to be less prone to blurred action).

Top of playback picture on TV waves back and forth. EVF display is good. This condition indicates a possible compatibility problem between the camcorder and TV. Camcorder signals are not as stable as an off-the-air TV signal. The symptom is usually more noticeably when a prerecorded tape, recorded by another camcorder or VCR, is played back. In extreme cases, it is necessary to alter the TV circuits. This is discussed in Sec. 9-3. Before going to such drastic measures, try correcting the problem by adjustment of the horizontal hold on the TV.

9-3 CAMCORDER TROUBLESHOOTING/REPAIR NOTES

The following notes summarize practical suggestions for troubleshooting all types of camcorders. Note that these suggestions supplement the troubleshooting procedures given at the end of circuit descriptions in Chapters 4 through 7.

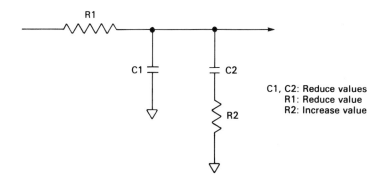

FIGURE 9-1 Altering the integrating circuits of a TV AFC

9-3.1 TV AFC Compatibility Problems

If the AFC circuits of a TV are not compatible with the VCR section of a camcorder, *skewing* may result. Generally, the term *skew* or *skewing* in camcorders and VCRs is applied when the upper part of the reproduced picture is bent or distorted by *incorrect back tension* on the tape. (Typical back-tension adjustments are described in Sec. 8-1.3.)

The same effect can be produced when the time constant of the TV AFC circuits cannot follow the camcorder/VCR playback output. This condition is *very rare* in newer TV sets (which are designed with VCRs, videodisc players, and camcorders in mind) and appears only about 1% of the older TV sets and almost never when the camcorder/VCR and TV are produced by the same manufacturer). So, *do not change* the TV AFC circuits unless you are *absolutely certain* there is a problem. At least, try the camcorder with a different TV *first*.

Once you are absolutely certain that there is a problem with compatibility, you can reduce the time constant of the *integrating circuit* associated with the TV AFC. Figure 9-1 shows the major components of a typical TV AFC integrating circuit. To reduce the time constant, reduce the values of either or both capacitors C1 and C2, reduce the value of R1, and increase the value of R2. It is generally not necessary to change all four values. Be sure to check the stability of the TV horizontal sync after changing any AFC value, since the AFC circuits are usually part of the horizontal sync system in modern TV sets.

9-3.2 Copy Problems

It is possible (but not recommended by the author) to copy a videocassette using two camcorders or a camcorder/VCR combination. One unit plays the cassette to be copied; the other unit makes the copy. Keep two points in mind when making such copies.

First, if the cassette being copied contains any copyrighted material, you may be doing something illegal! Second, a copy is never as good as the original, and copies of copies are usually terrible. Even with professional recording and copying equipment, the quality of a copy (particularly the color) deteriorates with each copying. The quality of a first copy (called the second generation) can be "acceptable," provided that the original is of very good quality. However, a second copy (third generation) is probably of unacceptable quality. Forget fourth-generation (or beyond) copies. So if you are called in to service a camcorder that "will not make good copies of other cassettes," explain that the problem probably has no cure.

9-3.3 Wow and Flutter Problems

Camcorders are subject to wow and flutter, as are most VCRs (and audio recorders). Wow and flutter are tape-transport speed fluctuations that may cause a regularly occurring instability in the picture and a quivering or wavering effect in the sound during record and playback. The longer fluctuations (below about 3 Hz) are called *wow*; shorter fluctuations (typically 3 to 20 Hz) are called *flutter*.

Wow and flutter can be caused by mechanical problems in the tape transport or by the servo system. Wow and flutter are almost always present in all camcorders, but it is only when the wow/flutter condition goes beyond a certain tolerance that the problem is objectionable.

If you are to service a camcorder where the complaint appears to be excessive wow and flutter, first check the actual amount. This can be done with a wow and flutter meter. However, it is generally more practical to use the low-frequency tone recorded on an alignment tape and a frequency counter connected to the audio line at some convenient point (at the A/V output connector, for example). Typically, the low-frequency tone is on the order of 333 Hz, and an acceptable tolerance is 0.03%. If necessary, it may be necessary to operate the frequency counter in the PERIOD mode to increase resolution (to read a value such as 0.0999 Hz).

9-3.4 System-Control Problems

It is difficult to generalize about system-control (Chapter 7) service problems. In camcorders, system control is performed by a microprocessor (also called a microcomputer in some literature). In the simplest of terms, you press front-panel buttons to initiate a given operating mode, and the microprocessor produces the necessary control signals (to operate relays, motors, etc.). Each camcorder has its own system-control functions. *You must learn these functions.*

System control for all camcorders has certain *automatic-stop* functions, such as end-of-tape stop, condensation detector (dew sensor), reel-rotation detector, and so on, which must be accounted for in service. For example, when checking any system control, make certain that all the automatic-stop functions are capable of working. Then make certain that some automatic-stop function has not worked at the wrong time (end-of-tape stop occurs in the middle of the tape, for example).

The automatic-stop functions can also be a source of imaginary trouble. For example, do not expect the tape to keep moving when the end-of-tape stop has occurred.

It is often necessary to disable the automatic-stop functions during service. The following notes describe some generalized procedures for checking and testing system-control functions.

To disable the end-of-tape sensor function at both ends of the tape, cover the *end sensor lamp* (such as D125 in Fig. 7-12) with opaque tape. As an alternative, cover the *takeup or supply-end sensors* (Q123/Q124 in Fig. 7-12).

If the camcorder has a switch that is actuated when a cassette is in place (such as cassette up/down switch S121 in Fig. 7-12), locate the mechanism that actuates the switch, and hold the mechanism in place with tape. The same applies to the safety tab switch S122. In many (but not all) cases, it is possible to operate the camcorder through all operating modes with a cassette installed if the cassette switches can be actuated manually.

Reel-rotation detectors (such as reel sensor Q119 in Fig. 7-12) can be checked by holding the reel. This causes the reel clutch to slip (to prevent damage), but the detector senses that the reel is not turning and produces automatic stop. Note that some camcorders have two reel-rotation detectors or sensors, one for takeup and one for supply.

Always check that all automatic-stop functions work and that all bypasses and simulations are removed after any service work!

9-3.5 Interchange Problems

When a camcorder can play back its own recordings with good quality, but playback of tapes recorded on other machines is poor, the camcorder is said to have interchange problems (the camcorder cannot interchange tapes). Such problems are always located in the VCR section of the camcorder, usually in the tape path. Quite often, interchange problems are the result of improper adjustment. We describe interchange adjustments in Secs. 8-1.9 and 8-1.10.

9-3.6 Servo System Problems

Total failures in the servo system (Chapter 5) are usually easy to find. If a servo motor fails to operate, check that power is applied to the motor at the appropriate time. If power is there, but the motor does not operate, the motor is at fault (burned out, open windings, etc.). If the power is absent, track the power line back to the source. (Is the microprocessor delivering the necessary control signal or power to the relay or IC?)

The problem is not so easy to locate when the servo fails to lock on either (or both) record and playback, or locks up at the wrong time (causing the heads to mistrack even slightly). Obviously, if the control signal is not recorded (or is improperly recorded) on the control track during record, the servo cannot lock properly during playback. So your first step is to see if the servo can play back a properly recorded tape.

There are usually some obvious symptoms when the servo is not locking properly. (There is a horizontal band of noise that moves vertically through the picture if the servo is out of sync during playback.) The picture may appear normal at times, possibly leading you to *think* that you have an intermittent condition. However, with a true out-of-sync condition, the noise band appears regularly, sometimes covering the entire picture.

Keep in mind that the out-of-sync condition during playback can be the result of servo failure or of the fact that the sync signals (control signals) are not properly recorded on the tape-control track. To find out if the servo is capable of locking properly, play back a known-good tape. If the playback is out of sync, you definitely have a servo problem.

The symptoms for failure of the servo to lock during record are about the same as during playback, with one major exception. During record, the head-switching point (which appears as a break in the horizontal noise band) appears to move vertically through the picture in a random fashion.

Another way to check if the servo is locking on either record or playback involves looking at some point on the rotating scanner (or cylinder) under fluorescent light. When the servo is locked, the fluorescent light produces a blurred pattern on the rotating scanner that appears almost stationary. When the servo is unlocked, the pattern appears to spin.

Try checking the scanner of a known-good camcorder under fluorescent light. Stop and start the camcorder in the RECORD mode. Note that the blurred pattern appears to spin when the scanner first starts, but settles down to almost stationary when the servo locks. Repeat this several times until you become familiar with the appearance of a locked and unlocked servo under fluorescent light.

Once you have studied the symptoms and checked the servo playback with a known-good tape, you can use the results to localize trouble in the servo. For example, if the servo remains locked during playback of a known-

Sec. 9-4   Trouble Symptoms Related to Adjustment   293

good tape, it is reasonable to assume that the circuits between the control head and servo motors are good. Note that such circuits include both speed and phase control, as is typical for modern VCRs.

Keep in mind that servo troubles may be mechanical or electrical and may be the result of improper adjustment or component failure, or both. As a general guideline, if you suspect a servo problem, make the electrical adjustments that apply to the servo. Always follow the manufacturer's adjustment procedures, using the procedures of Chapter 8 as a guide. This may cure the servo problem. If not, simply following the adjustment procedure tells you (at the very least) if all the servo-control signals are available at the appropriate points in the circuits. If one or more of the signals are found to be missing or abnormal during adjustment, you have an excellent starting point for troubleshooting.

9-4 TROUBLE SYMPTOMS RELATED TO ADJUSTMENT

The following notes describe trouble symptoms that can be caused by improper adjustment. Keep in mind that these same symptoms can be caused by circuit failures other than adjustment. (We describe circuit failure on the basis of trouble symptoms in the remaining sections of this chapter.)

Picture displaced in the line following head switching. This problem, often called *skew error* or *skew*, is often the result of improper back-tension adjustment (Sec. 8-1.3).

Tape stretched, broken, loose in cassette or spilled from cassette. These problems can usually be corrected by proper cleaning of the brakes. However, if the symptoms remain after cleaning, it may be necessary to replace the brakes.

Tape creased or frilled on the edges. These problems can usually be corrected by tape guide or other tape path adjustments described in Sec. 8-1.

No color, no color sync, or improper tint. Assuming a known-good monitor TV (properly fine tuned if necessary), these problems are usually the result of an improperly adjusted subcarrier oscillator (Sec. 8-2.2) but can also be caused by chroma carrier level adjustments (Sec. 8-2.14).

No picture on EVF or monitor TV during record. This can be caused by a missing or improperly adjusted target voltage (Sec. 8-2.6).

Image appears to be dark or light under normal lighting conditions. This condition can be caused by improper adjustment of the AIC level (Sec. 8-2.7).

Picture heavily shaded or washed out. This condition can be caused by improper adjustment of the beam current (Sec. 8-2.9).

Picture streaked in the horizontal direction. This condition can be caused by improper adjustment of the streaking control (Sec. 8-2.10).

Picture does not become totally dark, or there is color noise, when camera is aimed at a dark area. This condition is usually the result of improperly adjusted dark shading (Sec. 8-2.11) but can also be caused by dark-offset adjustments (Sec. 8-2.12) or chroma setup adjustments (Sec. 8-2.19).

Focus does not track throughout the entire zoom range. This problem is usually the result of improper backfocus adjustment (Sec. 8-2.13).

Distorted images or space at the top or bottom of the picture. This problem is usually the result of improper vertical size adjustment (Sec. 8-2.15).

Distorted images together with a shift in color balance. This problem is usually the result of improper horizontal size and linearity adjustment (Sec. 8-2.16).

Space on either side of the picture. This problem is usually the result of improper horizontal and/or vertical center adjustment (Sec. 8-2.17).

Colored edges in the line following a dark-to-light or light-to-dark transitions. This problem is usually the result of improper vertical-edge correction adjustment (Sec. 8-2.18).

Uneven red colors. This problem is usually the result of improper R-signal shading adjustment (Sec. 8-2.30).

Uneven blue colors. This problem is usually the result of improper B-signal shading adjustment (Sec. 8-2.31).

Red and blue colors change with changes in light levels. This problem is usually the result of improper tracking adjustment (Sec. 8-2.32).

EVF picture out of horizontal sync. This problem is usually the result of improper EVF horizontal hold adjustment (Sec. 8-2.38).

EVF picture tilted. This problem is usually the result of improper EVF picture tilt adjustments (Sec. 8-2.39).

EVF picture not centered. This problem is usually the result of improper EVF centering magnets adjustment (Sec. 8-2.40).

EVF picture distorted. This problem is usually the result of improper EVF vertical size adjustment (Sec. 8-2.41).

EVF picture out of focus. This problem is usually the result of improper EVF focus adjustment (Sec. 8-2.42).

EVF picture lacks contrast or brightness. This problem is usually the result of improper EVF contrast and brightness adjustment (Sec. 8-2.43).

No autofocus. Manual focus is good. This problem is usually the result of improper autofocus adjustments (Secs. 8-2.44 through 8-2.49).

No overdischarge indication. Usable battery time shortened. This problem can be the result of improper overdischarge-level adjustment (Sec. 8-2.51), but there can also be a problem with the battery.

Horizontal instability and/or picture noise. This problem can be the result of improper cylinder motor-speed adjustment (Sec. 8-2.52).

Sec. 9-6   No Picture on EVF or Monitor Power On, Camera Mode

Vertical instability and/or picture noise. This problem can be the result of improper capstan motor-speed adjustment (Sec. 8-2.53).

Horizontal and vertical instability. This problem can be the result of improper 30-Hz reference frequency adjustment (Sec. 8-2.54).

Vertical jitter (head-switching noise). This problem can be the result of improper PG shifter adjustment (Sec. 8-2.55).

Noise in picture, but picture stable. This problem is usually the result of improper tracking preset adjustment (Sec. 8-2.56), but can also be caused by the user tracking control.

Color degraded or diamond-shaped beats in picture. This problem is usually the result of improper record chroma-level adjustment (Sec. 8-2.57).

Picture "overloaded" or washed out. This problem can be the result of improper record luma level adjustment (Sec. 8-2.58).

Audio weak. This problem can be the result of improper audio playback gain (Sec. 8-2.59).

Audio distorted and/or high frequencies attenuated. This problem can be the result of improper audio bias-level adjustment (Sec. 8-2.60).

9-5   USING THE TROUBLESHOOTING PROCEDURES

The remainder of this chapter is devoted to step-by-step troubleshooting procedures for typical camcorders. The troubleshooting approach in this chapter is based on *trouble symptoms* (the most common troubles reported to manufacturing service personnel). A separate section is devoted to each symptom. These symptoms can apply to any camcorder, but are related specifically to the camcorders described in Chapters 4 through 7.

After selecting the symptom (or symptoms) which match(es) those of the camcorder being serviced, follow the steps in the troubleshooting procedure. Note that the simplified or block diagrams in Chapters 4 through 7 are referenced from the procedures in this chapter. The procedures here help isolate the problem to a defective module or component shown on the diagrams. If adjustments are involved in the procedures, direct reference is made to the corresponding procedure in Chapter 8.

9-6   NO PICTURE ON EVF OR MONITOR POWER ON, CAMERA MODE

Figures 6-1, 6-2, and 6-3 show the circuits involved. This symptom assumes that CAMERA mode is selected, there is a raster on both the EVF and monitor, but there is no picture on either display.

If there is a picture on the monitor, but not on the EVF, check for video at pin 7 of the EVF connector. If absent, suspect IC960. If present, suspect the EVF circuits (Sec. 9-16). Also check Q1514/Q1515.

If there is a picture on the EVF, but not on the monitor, check for video at pin 8 of IC960 or TP904. If absent, suspect IC960 and/or C909. Also check Q1514/Q1515.

Start by applying a colorbar signal (1 V p-p) to the video line (at pin 6 of the EVF connector). If you get a good colorbar picture on the monitor, suspect the camera circuits.

Next, check for video (1 V p-p) at pin 11 of IC203. If present, check for video at pins 3 and 8 of IC906. If absent, suspect IC960 and/or C917. Also check for video on both sides of C909/C911.

If video is absent at pin 11 of IC203, check for 5 V at pin 16 of IC203. If absent, check the PB 5 V line. Also check for video of about 220 mV p-p at pins 7 and 9 of IC203. If present at pin 9 but not at pin 7, suspect C240.

Finally, check for video at pin 2 of IC203 (about 10 V p-p). If absent, suspect C254 or the camera circuits ahead of C254.

9-7 NO RECORD VIDEO

Figures 6-2 and 6-3 show the circuits involved. This symptom assumes that CAMERA mode is selected, the camcorder is in record, the circuits check out as described in Sec. 9-6, but there is no video recorded on tape.

Start by checking for camera video of about 4 V p-p at pin 13 of IC202. If present, check for an FM signal of about 150 mV p-p at TP201. If present, suspect the cassette tape, video heads, or IC201. Also check for 5 V at pin 2 of IC201. If absent, check the REC 5-V line.

Next check for a luma FM signal of about 350 mV p-p at pin 28 of IC202. If present, check the monitor cut/load line at pin 24 of IC202. This line should be low during record. If not, suspect system control IC901.

If the luma FM signal is absent at pin 28 of IC202, check for a luma FM signal of about 650 mV p-p at pin 17 of IC203. If present at IC203-17 but absent at IC202-28, suspect RT201 or the high-pass filter between IC203 and IC202.

If the luma FM signal is absent at IC203-17, suspect IC203. However, before pulling IC203, check for video at the following points: IC203-3 220 mV p-p, IC203-23 200 mV p-p, IC203-27 340 mV p-p, and IC203-25 300 mV p-p. If any of these signals is absent or abnormal, check the corresponding components L213, C243, C235, CP202, and Q273).

9-8 NO PLAYBACK VIDEO

Figure 6-3 shows the circuits involved. This symptom assumes that PLAYBACK mode is selected, the circuits check out as described in Sec. 9-6, but there is no video played back from a known-good tape.

Sec. 9-9 No Power to Camera with EVF Raster On

Start by checking for video of about 640 mV p-p at pin 22 of IC202. If absent, suspect IC202, IC201, or the video heads. However, before pulling these components, check for signals at the following points: IC202-18 9 V, IC202-16 40 mV p-p, IC202-25 120 mV p-p, IC 201-1 15 Hz, and IC202-7 30 Hz.

If IC202 is not 9 V, check the PB 9-V line.

If IC202-16 is not 40 mV p-p, but IC202-25 is 120 mV p-p, suspect CP201. If IC202-25 is not 120 mV p-p, suspect IC202.

If IC202-1 is not 15 Hz, check the SW 15-Hz line. If IC202-7 is not 30 Hz, check the SW 30-Hz line.

If there is video of about 640 mV p-p at pin 22 of IC202, but no playback, suspect IC203. Before pulling IC203, check the following: monitor cut/load line at IC203-12 should be low, the playback line at IC203-16 should be high, IC203-26 should be 340 mV p-p, IC203-27 should be 230 mV p-p, and IC203-20 should be 560 mV p-p.

If IC203-12 is not low, suspect system control IC901. If IC203-16 is not 9 V, check the PB 9-V line.

If IC203-26 is not 340 mV p-p but IC203-27 is 230 mV p-p, suspect CP202, Q207/Q208, and C238. If IC203-27 is not 230 mV p-p, suspect IC203.

If IC203-20 is not 560 mV p-p but pin 22 of IC202 is 640 mV p-p, suspect Q204/Q205 and C224.

9-9 NO POWER TO CAMERA WITH EVF RASTER ON

Figure 4-4 shows the circuits involved. This symptom assumes that there is an EVF raster, the CAMERA and NORMAL modes are selected (stand-by switch on), but there is no power to the camera circuits. (This is not to be confused with a true "no-power" symptom such as described in Sec. 9-24.)

Start by checking for about 8.4 V at pin 3 of IC1604. If absent, check pin 4 of IC1604. If pin 4 is 0 V, check Q1507/Q1510. If pin 4 is 9.6 V, but pin 3 is not 8.4 V, suspect IC1604.

Next check for about 8.3 V at pin 6 of PG1507. If absent, suspect S1602.

Check for about 5.3 V at PG1312-5. If absent, suspect IC1305 and Q1317/Q1318.

Check for about 5.1 V at pin 12 of CN1509. If absent, check for about 4.9 V at pin 8 of PG901. If absent, check the $\overline{\text{STBY}}$ 5-V line. If present, check the base of Q1502. If the base is about 4.4 V, suspect Q1502. If the base is 4.8 V suspect Q1505/D1502.

9-10 NO-RECORD CHROMA

Figures 6-2 and 6-3 show the circuits involved. This symptom assumes that there is a picture but no color recorded on tape.

Start by checking for chroma signals at pin 26 of IC202. If present, but no color is being recorded, suspect IC202.

Next check for about 65 mV p-p at pin 20 of IC204. If the signal is present, but not at 65–80 mV p-p, suspect RT202.

If there is no signal at pin 20 of IC204, check for about 350 mV p-p at pin 14 of IC204 and about 260 mV p-p at pin 26 of IC204. If the signal is present at pin 14 but not at pin 26, suspect CP204.

If the signal is absent at pin 14, suspect IC204. Also check that pin 12 is at 0 V. If not, check the PB 5-V line.

9-11 NO PLAYBACK CHROMA

Figures 6-2 and 6-3 show the circuits involved. This symptom assumes that there is a picture but no color played back from a known-good tape.

Start by checking for chroma of about 130 mV p-p at pin 10 of IC203. If present, suspect IC203.

Next check for chroma of about 600 mV p-p at pin 14 of IC204. If absent, check that pin 12 of IC204 is at 5 V. If so, suspect IC204. If absent, check the PB 5-V line.

If there is chroma at pin 14 of IC204, check for chroma of about 540 mV p-p at pin 26, 1 V p-p at pin 24, and 200 mV p-p at pin 9.

If the signal is absent at pin 26, suspect CP204. If the signal is absent at pin 24, suspect IC204. If the signal is absent at pin 9, suspect DL202.

9-12 POWER ZOOM DOES NOT OPERATE

Figure 4-28 shows the circuits involved. This symptom assumes that the power zoom function is totally inoperative, but all other functions are normal.

Start by checking for about 3.6 V at pin 1 of PG1520. If absent suspect Q1501/ZD1501.

Next press and hold the telephoto (S1602) and wide-angle (S1603) switches, in turn, while checking the voltages at pins 1 and 2 of PG1621.

When S1603 is pressed, pin 1 should be about 2.7 V, with pin 2 at 0 V. If not, suspect S1605. When S1602 is pressed, pin 1 should be 0 V, with pin 2 at 2.7 V. If not, suspect S1602.

If the voltages are good at both pins but there is no power zoom function, suspect the zoom motor.

9-13 VIDEO LEVEL TOO HIGH OR LOW

Figures 4-11, 4-13, and 4-19 show the circuits involved. This symptom assumes that the picture is either "overloaded" or "washed out," with normal lighting. A malfunction in either the AIC or AGC circuits can cause such a condition.

Start by checking for video of about 640 mV p-p at pin 23 of IC1101 and about 440 mV p-p at pin 20.

If the video is good at pin 23 but not at pin 20, suspect IC1101. If the video is incorrect at pin 23, suspect Q1137/C1103.

Next check for a voltage of about 2.3 V at pin 18 of IC1101. If absent, suspect IC1502.

Next rotate AIC control RT1102, and check that the voltage at pin 14 of IC1101 changes. If not, check for about 4.4 V at pin 3 of PG1115. If present, suspect IC1101.

If the voltage is absent at pin 3 of PG1115, check the voltage at the base of Q1307. If the voltage is zero, suspect Q1306, C1317, and R1343. If the base voltage of Q1307 is about 0.6 V, but the voltage at pin 3 of PG1115 is absent or abnormal, suspect Q1307.

Next check that the voltage at IC1103-7 changes when RT1102 is rotated. If not, suspect IC1103.

Check the voltage at pin 1 of PG1118 (the drive voltage to the iris motor). This voltage should vary from about 0 V to 3.5 V during normal operation and/or when RT1102 is varied. If the voltage varies, but there is no change in the picture (remains overloaded or washed out), suspect the iris motor. If the voltage does not vary, suspect Q1135/Q1136.

9-14 NO PICTURE

Figures 4–7, 4–9, 4–10, 4–15, 4–17, 4–18, 4–19, 4–20, 4–26, and 6-1 show the circuits involved. This symptom assumes that the camcorder has been checked as described in Sec. 9-6, but there is no picture. The CAMERA and NORMAL modes are selected.

Start by checking for a prevideo signal of about 520 mV p-p at TP1101. If there is no prevideo, check for 8 V at pin 2 of T1301. If the 8V is absent, suspect IC1305, Q1310, and Q1311. Also check the power circuits as described in Sec. 9-9.

If T1301-2 is 8 V, but there is no prevideo, check for a flyback pulse of about 140 V p-p at pin 1 of T1301. If flyback is missing, check for a WHD pulse of about 4.8 V p-p at PG1115-4. If present, suspect Q1308, Q1309, C1318, and D1303. If absent, suspect IC1106 and Q1145.

If the flyback pulses are present, check the Saticon high voltages. If absent or abnormal, suspect T1301.

If all Saticon voltages are good, check for a horizontal deflection voltage of about 2 V p-p at pin 4 or PG1311. If absent, check for a CP2 pulse of about 4.8 V p-p at pin 1 of PG1115. If present, suspect IC1303, Q1301, Q1302, and Q1321. If absent, suspect IC1106 and IC1147.

If the horizontal deflection is good, check for a vertical deflection voltage of about 6 V p-p at pin 2 of PG1311. If absent, check for a VP pulse of about 4.8 V p-p at pin 2 of PG1115. If present, suspect IC1304 and Q1303. If absent, suspect IC1106.

Next check that noise appears when the target lead is touched. If not, suspect Q1001, Q1002, Q1004, Q1005, and Q1006. If noise appears, check that the iris is open and that the AIC circuits are operating normally as described in Sec. 9-13.

If the iris is open, all the voltages to the Saticon are good, and noise appears when the target lead is touched, but there is no picture or no signal to TP1101, suspect the Saticon.

If there is a prevideo signal at TP1101, but no picture, check for a luma signal of about 200 mV p-p at TP1104. If absent, check for luma of about 560 mV p-p at TP1103. If a good signal is present at TP1103, but not at TP1104, suspect CP1103, DL1101, and Q1106.

If the signal at TP1103 is absent or abnormal, check for prevideo of about 440 mV p-p at pin 27 of IC1101. If absent, suspect IC1101. If present, suspect Q1101 through Q1104.

If the signal at TP1104 is normal, check for a luma signal of about 780 mV p-p at pin 13 of IC1102. If absent, suspect IC1102.

If the signal at IC1102-13 is good, check for a luma signal of about 600 mV p-p at pin 25 of IC1105. If absent, suspect Q1117 and Q1109.

If the signal at IC1105-25 is normal, check for a video signal of about 1.8 V p-p at pin 27 of IC1105. If absent, check that pin 13 of IC1105 is zero. If not, check the inhibit line. Also check for a C SYNC signal at pin 28, a C BLK signal at pin 20, and a BGP signal at pin 18 of IC1105. If any of these signals is absent or abnormal, suspect IC1106 and Q1146.

If the signal at pin 27 of IC1106 is normal, check for a video signal of about 1 V p-p at TP1107 or PG1109-10. If absent, suspect Q1134 and Q1133.

If the signal at PG1109-10/TP1107 is normal, check for video at CN1501-13. If absent, suspect Q1514 and D1502. The Q1514 gate should be about 3.8 V. If not, check the C STBY 5-V line and Q1515. (Q1515 should be on only when the external character generator is used.)

9-15 NO AUTOFOCUS

Figures 4-29, 4-30, and 4-31 show the circuits involved. This symptom assumes that the camcorder can be focused manually, but there is no automatic focus when the focus switch is set to "auto."

Sec. 9-16   No Color

First, make certain that the focus switch S3 is set to auto and that C STBY 12 V and C STBY 5 V are available. Check for C STBY 5 V at pin 3 of IC5. If absent, check for C STBY 12 V at pin 1 of IC5. If absent, suspect Q13 and Q14.

Check for 3 V (V-reference) at pin 16 of IC1. If absent or abnormal, suspect IC1.

Check for signals at TL1 and TL2. If absent, check for signals at pins 1 and 20 of IC1. If absent, check for a drive signal to infrared LED D6. If absent, suspect Q11, Q12, Q15, Q17, and IC2. If there is drive to D6, but no signals at pins 1 and 20 of IC1, suspect D4 and D6. (Also make sure the lens cap is off!)

Note that the drive signal from IC2 to D6 is controlled by the IC2 oscillator, which, in turn, is controlled by SL-A and SL-B. The oscillator frequency should be about 833 kHz and produce sync signals of about 110-μs duration at pin 18 of IC2.

If there are signals at TL1 and TL2, check for signals at pins 2 through 5 of IC2. If absent, suspect IC3. If present, check for signals (far and near signals) at pins 6 and 7 of IC2. If absent, suspect IC2.

If the far and near signals are correct (as shown by the truth table of Fig. 4-31), but the focus motor does not respond properly, suspect Q3 through Q10 and Q16. Also check adjustment of RT5, and determine that far switch S1 is properly actuated.

Before pulling IC1, make certain that the clear and sync signals (at pins 7 and 8 of IC1, respectively) are present. If not, suspect Q1 and Q2. Check for inverted clear and sync signals at pins 16 and 18 of IC2. Also check adjustment of RT1 (gain), RT2 (B-channel offset), and RT3 (A-channel offset). Finally, if the TL1/TL2 signals are absent or abnormal, but all signals to IC1 are correct, and all IC1 adjustments have been made, check C15 and C16.

9-16 NO COLOR

Figures 4-19, 4-21, 4-24, 4-25, and 4-26 show the circuits involved. This symptom assumes that there is a picture, but no color, on a known-good monitor TV and that the circuits have been checked as described in Sec. 9-14.

Start by checking for R and B color signals of about 150 mV at TP1109 and TP1110. If either signal is absent, check for Rc/Bc-signals of about 50 mV p-p at IC1104-36. If absent or abnormal, suspect CP1105, and check adjustment of RT1104.

If the signal is good at IC1104-36, check for Rc/Bc-signals of about 700 mV p-p at TP1108. If absent, suspect IC1104.

If the signal is good at TP1108, check for Rc/Bc-signals of about 50 mV p-p at pin 33 of IC1104. If absent or abnormal, suspect DL1103, and check adjustment of RT1105.

If the signal is good at IC1104-33, check for Rc/Bc-signals of about 200 mV p-p at pin 37 of IC1104. If absent or abnormal, suspect Q1121 and Q1122, and check adjustment of RT1106.

If the signal is good at IC1104-37, check for R- and B-signals of about 310 mV p-p at pin 16 (and 300 mV p-p at pin 4) of IC1104. If absent at either pin, suspect IC1104.

If the signals are good at pins 4 through 16 of IC1104, check for R- and B-signals of about 200 mV p-p at pin 6 (and 80 mV p-p at pin 14) of IC1104. If absent, check adjustment of RT1113 and RT1114.

Check for R- and B-signals of about 200 mV p-p at TP1109 and TP1110. If absent, suspect IC1104. If the signals are good at TP1109 and TP1110, check for a YL-signal of about 150 mV p-p at TP1105. If absent, suspect Q1105 and CP1102. (Note that if there is a picture, the signal at TP1103 should be good, as described in Sec. 9-14).

If the YL-signal is good at TP1105, check for R-YL-signals of about 560 mV p-p and B-YL-signals of about 340 mV p-p at TP1113 and TP1114, respectively. If absent, suspect IC1104.

Next check for SC1 and SC2 signals of about 0.5 to 0.7 V p-p at pins 23 and 8 of IC1105. If absent, suspect IC1106.

Check for chroma signals of about 640 mV p-p at pin 4 of IC1105. If absent, check an inhibit signal of 5 V at IC1105-13, a chroma clip signal at IC1105-14, and a vertical edge-suppress signal at IC1105-15. If the inhibit signal is absent or abnormal, suspect system control IC901. If the chroma clip signal is absent, suspect DL1104. If the vertical edge suppress signal is absent, suspect Q1131.

If the chroma signal is good at IC1105-4, check for a chroma signal of about 100 mV at pin 3 of IC1105. If absent or abnormal, suspect DL1105 and check adjustment or RT1120.

If the chroma signal at IC1105-3 is good, check for a chroma signal of about 380 mV p-p at pin 5 of IC1105. If present, but there is no color, suspect L1112. If absent, check for a burst signal of about 0.5 V p-p at TP1115. If absent, suspect IC1106, RT1121, Q1143, and Q1144. Also check for a BF-signal of about 2.8 V at pin 11 of IC1105. If absent, suspect IC1106. If the burst (pin 1), BF (pin 11), and chroma (pin 3) signals are good, but there is no chroma at pin 5, suspect IC1105.

9-17 INCORRECT COLOR SHADING

Figures 4-11, 4-12, 4-15, 4-18, and 4-23 show the circuits involved. This symptom assumes that there is a color picture but that color shading is incorrect.

Start by checking for a prevideo signal of about 520 mV p-p at TP1101. If necessary, readjust the beam current with RT1324.

Check for a dynamic focus correction signal of about 800 mV p-p at TP1503. If absent, check for signals of about 50 mV p-p at IC1503-2, 160 mV p-p at IC1503-3, 50 mV p-p at IC1503-4/5, 1.2 V p-p at IC1502-2, 0.7 V p-p at IC1502-3, 0.8 V p-p at IC1502-7, and 1.3 V p-p at IC1502-9.

If the signal is present at IC1503-2, but not at TP1503, suspect IC1503. If the signal is present at IC1503-3, but not at IC1503-2, suspect C1509. If the signal is absent at IC1503-3, but are present at IC1503-4/5, suspect IC1503. If the signals at IC1503-4/5 are absent or abnormal, suspect IC1502, and check adjustment of RT1505, RT1506, RT1507, and RT1508.

If the signal is good at TP1503, check for a horizontal dark-shading correction signal of about 0.2 V p-p at TP1501. If absent or abnormal, suspect IC1503, and check adjustment of RT1501 and RT1502.

If the signal is good at TP1501, check for a vertical dark-shading correction signal of about 70 mV p-p at TP1502. If absent or abnormal, suspect IC1503, and check adjustment of RT1503 and RT1504.

If the signal is good at TP1502, check for a blue color-shading correction signal of about 0.6 V p-p at TP1504. If absent or abnormal, suspect IC1503, and check adjustment of RT1509 through RT1512.

If the signal is good at TP1504, check for a red color-shading correction signal of about 0.96 V p-p at TP1505. If absent or abnormal, suspect IC1503, and check adjustment of RT1513 through RT1516.

If all the circuits and adjustments described thus far appear to be good, but color shading is incorrect, try adjustment of the Saticon focus. Then check the Saticon tube and the Saticon deflection yoke. Improper focus and/or deflection problems can cause incorrect color shading.

9-18 INCORRECT WHITE BALANCE (INCORRECT COLOR BALANCE)

Figures 4-24 and 4-27 show the circuits involved. This symptom assumes that color balance is incorrect, and it is suspected that white-balance circuits may be the cause. (If color balance is good, leave the white balance alone!)

Start by setting the white-balance switch S4 to manual and checking if color balance can be changed by adjustment of balance control RT1601. If not, check that the voltage at IC1104-32 varies from about 1.6 to 1.9 V, when S1601 is rotated.

If the voltage at IC1104-32 varies, but there is no change in color balance, suspect IC1104. If the voltage does not change at IC1104-32, suspect D1105. (It is also possible that Q1140 and Q1141 have been turned on, cutting RT1601 out of the circuit.)

If RT1601 appears to have control over color balance, rotate RT1601 to either extreme (unbalance the color), and set S4 to auto. Check that color balance is returned when S4 is set to auto. If there is a change, but color balance is still not good, it is possible that the R and B gain is not properly adjusted. Try readjustment of RT1113 and RT1114.

If color balance cannot be established with proper R and B gain, suspect IC1104. Before you pull IC1104, make the following checks. Check for an R-YL-signal of about 560 mV p-p at TP1113, a B-YL-signal of about 340 mV p-p at TP1114, a reset pulse at IC1104-20, a VP pulse of about 1.8 V p-p at IC1104-21, a CP1 pulse of about 3.4 V p-p at IC1104-15, and for 5 V at IC1104-13.

If the R-YL/B-YL-signals are absent, suspect IC1102. If the reset pulse is absent, suspect Q1506, D1503, Q1511, and Q1512. Note that the reset pulse (a momentary pulse) is applied when power is first turned on, and when S4 is changed from manual to auto.

If the VP pulse is absent, suspect IC1106. If the CP1 pulse is absent, suspect IC1106. If the 5 V is absent from IC1104-13, check the wiring from S4. (Note that this same 5 V turns Q1140/Q1141 on to remove RT1601 from the circuit.)

9-19 NO EVF RASTER

Figure 4-32 shows the circuits involved. This symptom assumes that there is power to the camera, but no EVF raster.

Start by checking for 9 V at pin 1 of T1801. If missing, check the 9-V line.

Next check for voltages from T1801 to the EVF CRT as shown in Fig. 4-32. If all the voltages are absent, check for a flyback pulse from Q1803. If one or more of the voltages is absent, check the corresponding circuit. For example, if there is no 2.6-kV power to the CRT, with a good flyback pulse, suspect T1801.

If all voltages from T1801 to the CRT are good, but there is no raster (or dot or trace), suspect the CRT.

If there is a horizontal trace, but no vertical deflection, check for vertical signals from IC1801 to the CRT deflection yoke. The vertical drive signal at IC1801-16 is about 7 V. If absent, suspect IC1801. If present, but no vertical deflection, suspect C1838, C1837, RT1802, and the deflection yoke.

Note that if there is no horizontal drive signal at IC1801-2, there will be no flyback pulse at T1801 (and no voltages to the CRT).

9-20 NO EVF PICTURE

Figures 4-32 and 6-1 show the circuits involved. This symptom assumes that there is an EVF raster, and there is video on the monitor TV, but no video on the EVF.

Sec. 9-21   No Sound (Audio)   **305**

Start by checking for a video signal of about 1.5 V p-p at IC1801-10, 300 mV p-p at IC1801-9, and 900 mV p-p at IC1801-11. If video is present at IC1801-9, but absent at IC1801-10, suspect IC1801. If video is present at IC1801-9, but absent at IC1801-11, suspect C1835, R1813, and C1814. If video is present at IC1801-11, but absent at IC1801-9, suspect R1804 and C1813.

If video is absent at all three points, trace the video back to the EVF connector (Fig. 6-1).

Finally, check for video of about 20 V p-p at the CRT grid. If present, but there is no video on the EVF screen, suspect the CRT. If there is no video at the CRT grid, but video at IC1801-10, suspect C1807, Q1802 and C1821.

9-21 NO SOUND (AUDIO)

Figures 6-7 through 6-10 show the circuits involved. This symptom assumes that there is no sound on the earphone or monitor TV during either playback or record, with all other camcorder functions normal.

Start by selecting the CAMERA mode and applying audio (1 kHz, 0.89 V p-p) to the EVF connector using an A/V input adapter.

If there is audio at either the earphone or the monitor TV, the problem is likely in the wiring. For example, if there is no audio at the earphone jack, but good audio at the monitor TV, check the earphone jack JK401 and corresponding wiring. If the audio is good at the earphone but not at the TV, check the wiring from pin 10 of IC401 to the TV (including the RF modulator or converter, if used).

If there is response at both the earphone and monitor TV when audio is applied to the EVF connector, but not when the microphone is used, check the wiring from pin 23 of IC401 to the microphone.

If there is no response at either the earphone or TV when audio is applied at the EVF connector, check that IC401 is in the CAMERA and RECORD mode rather than the PLAYBACK mode. Pin 2 of IC401 should be low (0 V) in record. Also note that if audio is good in PLAYBACK, but not RECORD, proceed as described in Sec. 9-23. If audio is good in RECORD, but not PLAYBACK, proceed as described in Sec. 9-22.

Next check for audio of about 350 mV p-p at pin 10 of IC401. If present, but there is no audio to the earphone or monitor TV, suspect C419. If absent at IC401-10, check for audio of about 65 mV p-p at pin 7 of IC401. If present, suspect IC401.

If audio is absent at IC401-7, check for audio of about 150 mV p-p at pin 13 of IC401. If present, but there is no audio at IC401-7, suspect the 15.75-kHz trap (C410/L401).

If absent at IC401-13, check for audio of about 70 mV p-p at pin 14 of IC401. If present, but there is no audio at IC401-13, suspect IC401.

If absent at IC401-14, check for audio at pin 22 of IC401. If present, but there is no audio at IC401-14, suspect C421, C424, C425, and R422. Also check that Q402 is not on. (Q402 is turned on by IC901 during playback to cut the microphone input.) If the base of Q402 is not low, check the microphone cut line (IC901-60).

9-22 NO RECORD AUDIO

Figures 6-7 through 6-10 show the circuits involved. This symptom assumes that the circuits are found to be good when checked as described in Sec. 9-21, there is no audio recorded on tape by the camcorder, but there is good audio playback from a prerecorded tape.

Start by checking for record audio of about 60 mV p-p at pin 5 of IC401. If absent, but there is audio at IC401-7, as described in Sec. 9-21, suspect C413.

Next check for audio of about 940 mV p-p at pin 4 of IC401. If audio is absent at pin 4, but present at pin 5 of IC401 in record, suspect IC401 (or possibly the record equalization network at pins 4 and 6 of IC401).

Next check for proper audio bias as described in Chapter 8. If the bias is not correct, and cannot be adjusted to the correct level, suspect Q404, and T401 (or possibly the erase heads). Also check that the oscillator-on line from IC901-59 is high (5 V) during record.

If the bias is good, and there is a good record audio signal at IC401-4, but no audio recorded on tape, check that TP402 is grounded. If not, suspect Q401 and IC402. Also check that the PB line from IC901-56 is low (the base of Q401 and pin 7 of IC402 low, pin 1 of IC402 not grounded).

If TP402 is grounded, IC402-1 is high, bias is good, and there is a good recording signal (about 940 mV p-p) at IC401-4, but no audio recorded on tape, suspect the audio record/playback head.

9-23 NO PLAYBACK AUDIO

Figures 6-7 through 6-10 show the circuits involved. This symptom assumes that the circuits are found to be good when checked as discussed in Sec. 9-21, but there is no audio when a known-good tape is played back (camcorder in PLAYBACK mode).

Start by checking for playback audio of about 5 mV p-p at pin 18 of IC401. If absent, check for audio at TP402. If present, at TP402, but absent at IC101-18, suspect C427.

If there is no playback audio at TP402, check that PG415-3 and IC402-1 are grounded. If not, suspect Q401, IC402. Also check that the PB line from IC901-56 is high (the base of Q401 and pin 7 of IC402 high, pin 1 of IC402 grounded).

If IC401-1 is grounded, but there is no playback audio at TP402 (from a known-good tape), suspect the audio record/playback head.

Next check for audio of about 300 mV p-p at pin 17 of IC401. If absent at pin 17, with good audio at pin 18, suspect IC401.

Next check for audio of about 120 mV p-p at pin 15 of IC401. If absent at pin 15, but with good audio at pin 17 of IC401, suspect RT401 (or possibly the equalization network at pins 17 and 19 of IC401).

Next check for audio of about 200 mV p-p at pin 13 of IC401. If absent at pin 13, but present at pin 15, suspect IC401.

Also check that pin 2 of IC401 is high and that pin 3 is low during playback. If not, suspect IC901.

9-24 NO POWER

Figure 7-5 shows the circuits involved. This symptom assumes that there is no power to any of the camcorder circuits.

Start by connecting a battery or adapter/charger and checking for 12 V at PG909-2. If absent, or below 10.9 V, suspect the battery or adapter/charger (or possibly J950).

Next check for 12 V at RL901. If absent, suspect F970.

Press and hold power switch S056. Check the voltage at the collector of Q904. If 0 V, check at the base of Q904. If the base of Q904 is 11.4 V, but the collector is zero, suspect Q904.

If the base of Q904 is 12 V, check the voltage at pin 12 of IC901. If zero, suspect ZD901. If 5 V, suspect D910 and S056.

Next check for 5 V at pin 26 of IC901. If zero after S056 is pressed, suspect IC905.

Next check at pin 43 of IC901. There should be a momentary (3-ms) reset pulse of 5 V; then pin 43 should return to zero, while pin 26 remains at 5 V). If there is no reset, suspect the reset circuit (Q903, ZD902, C903).

Now release S056, and check that pin 26 of IC901 remains at 5 V. If not, check the output from RL901. If the output is about 12 V, but IC901-26 does not remain at 5 V, suspect D902.

If the output from RL901 is 0 V, check that pin 52 of IC901 goes high for about 100 ms after the reset pulse at pin 43. If not, suspect IC901. If IC901-52 does go high for about 100 ms, but there is no output from RL901, suspect Q901 and RL901.

If pin 26 of IC901 remains at 5 V after S056 is released, check that 5 V is applied to all the regulator circuits shown in Figs. 7-5 through 7-7. Also check that each regulator produces the corresponding output.

Check the output of the switching regulator at PG901-8. If the output is 0 V, instead of 5 V, it is possible that the standby/normal switch (or power-save switch) is set to standby or power-save.

If the standby/normal switch is in NORMAL, but PG901-8 is not 5 V, suspect IC906 and Q951–Q954. Also check that pin 51 of IC901 is low. If IC901-51 is high, the switching regulator is turned off. Check that the base of Q1601 is low and that pin 9 of IC901 is not grounded.

9-25 BATTERY OVERDISCHARGE IS NOT DETECTED

Figure 7-8 shows the circuits involved. This symptom assumes that the battery discharge LEDs (or EVF displays) do not indicate when the battery is discharged.

Start by applying a variable power source to the external battery jack J950. Set the variable source to 12.3 ± 0.2 V. Check that the battery-F (full) LED D1601 turns off but that D1602 and D1603 remain on.

If D1601 does not turn off, check adjustment of ODC control RT901.

Next check for an anode voltage of 3.3 ± 0.2 V. If the anode voltage is absent or abnormal, suspect ZD901.

Check for a voltage of 1.6 ± 0.2 V at pin 7 of IC904. If the voltage is absent or abnormal, suspect RT901, R901, R902, and C920.

Check for 5 V at pin 9 of IC904. If absent, check the 5-V line. If present, and the voltage at pin 7 of IC904 is correct, but the battery discharge display is not correct (D1601 on, D1602/D1603 off), suspect IC904.

Reduce the variable power source at J950 to 11.6 ± 0.2 V. Check that both D1601 and D1602 are off, with D1603 on.

Reduce the variable power source at J950 to 10.9 ± 0.2 V. Check that all three LEDs D1601–D1603 turn off. (Also, check that the camcorder goes into the STOP mode and turns off, although the cassette can still be ejected.)

If the LEDs does not turn on and off at the correct voltages, and the condition cannot be corrected by adjustment of RT901, suspect IC904. Check the outputs at pins 2, 3, and 4 of IC904 against the truth table on Fig. 7-8.

If the LEDs turn on and off at the correct voltages, but the camcorder does not shut down when all three LEDs are off (or shuts down while one or more of the LEDs are on) suspect IC901.

9-26 FUNCTION SWITCHES INOPERATIVE OR MALFUNCTIONING

Figure 7-11 shows the circuits involved. If any or all of the function switches (camera and VCR operating controls) are inoperative or malfunctioning, follow the troubleshooting procedure described in Sec. 7-6.

Sec. 9-30   Mode Indicator Displays Inoperative or Malfunctioning   309

9-27 TROUBLE DETECTION CIRCUITS INOPERATIVE OR MALFUNCTIONING

Figure 7-12 shows the circuits involved. If any or all the trouble-detection functions are inoperative or malfunctioning, follow the troubleshooting procedures described in Sec. 7-7.

9-28 TAPE COUNTER INOPERATIVE

Figures 7-13 and 7-14 shows the circuits involved. If the tape-count display (on the LDC counter or the EVF) is not correct, follow the troubleshooting procedures described in Sec. 7-9.

9-29 CASSETTE DOES NOT EJECT

Figures 7-5 and 7-16 show the circuits involved. This symptom assumes that the cassette cannot be ejected by any means.

Start by checking that power is available as described in Sec. 9-24. Next, check if the cassette can be ejected in either the POWER-ON or POWER-OFF modes, but not both. If the cassette is ejected during POWER-ON, but not POWER-OFF, suspect D909.

Next press eject switch S057, and check the voltage at pin 50 of IC901. If 5 V, suspect D909/S057.

If IC901-50 drops to about 0.6 V when S057 is pressed, listen for sounds of the loading motor. If the motor appears to be running, but the cassette is not ejected, suspect the mechanism (Sec. 5-2).

If the loading motor does not run when S057 is pressed, check for 0 V at pin 54 of IC901 and 5 V at pin 55 of IC901. If absent or abnormal, suspect IC901.

Check for 9 V at pin 7 of IC903 and 0 V at pin 3 of IC903. If absent or abnormal, suspect IC903. If the voltages are correct at pins 3 and 7 of IC903, but the loading motor does not run, suspect the motor.

It is also possible that the mechanism-state switch is not providing proper data to IC901. For example, pins 39 and 40 should be high, and pin 41 should be low, during eject (as shown by the truth tables on Fig. 7-16).

9-30 MODE INDICATOR DISPLAYS INOPERATIVE OR MALFUNCTIONING

Figure 7-15 shows the circuits involved. If any or all of the mode indicator displays are inoperative or malfunctioning, follow the troubleshooting procedures described in Sec. 7-9.

9-31 HORIZONTAL STRIPES ON DISPLAY

Figure 5-9 and 5-12 show the circuits involved. This symptom assumes that horizontal stripes appear on both the EVF and monitor TV displays.

Start by checking adjustment of cylinder speed control RT604, as described in Chapter 8.

If this does not cure the problem, check for pulses of about 4.5 V p-p at pin 6 of IC602. If absent or abnormal, check for cylinder FG pulses of about 100 mV p-p at pins 2 and 3 of IC602. If absent or abnormal, suspect the cylinder motor.

Next check for cylinder FG pulses of about 2.6 V p-p at pins 4 and 5 of IC602. If absent at pin 4, but present at pins 2 and 3, suspect IC602. If absent at pin 5, but present at pin 4, suspect R651/C624. If cylinder FG pulses are present at pin 5 of IC602, but not at pin 6, suspect IC602.

If the pulses are good at pin 6 of IC602, check for a sawtooth signal at pin 12 of IC602. If absent, suspect IC602/C631.

If the sawtooth is good at pins 12 and 13 of IC602, check for about 2.5 V at pin 15 of IC602. If absent or abnormal, suspect IC602.

If the voltage is good at pin 15 of IC602, but the horizontal stripes remain, suspect IC607, R640/C648, and IC605.

9-32 CYLINDER DOES NOT ROTATE

Figures 5-12 and 5-14 show the circuits involved. This symptom assumes that the cylinder does not rotate in any mode.

Start by checking that power is available as described in Sec. 9-24. Make certain that the camcorder is not in STOP, REWIND, or FAST FORWARD modes (or any mode where the cylinder is not supposed to rotate).

With the camcorder in either playback or record, check for about 2.5 V at pin 2 of IC605. If present, check for about 12 V at pin 12 of IC605. If absent or abnormal, check the 12-V line. Also check for 5 V (Hall bias) at pin 9 of PG613. If absent, check the 5-V line.

If all voltages to IC605 are good, but the cylinder does not rotate, suspect IC605 or the cylinder motor.

If the voltage at pin 2 of IC605 is absent or abnormal, check for about 2.5 V at pin 3 of IC607 and pin 15 of IC602. If present at IC607-3, but not at IC605-2, suspect IC607, C648, and R640. If present at IC602-15, but not at IC607-3, suspect R635.

If the voltage is zero at pin 3 of IC607, check that the cylinder-on line and pin 7 of IC602 are both zero. If the cylinder-on line is 5 V in playback or record, check the line from IC901. If the cylinder-on line is zero, but pin 7 is high, suspect Q602.

Sec. 9-34   Picture Swings Horizontally, or Alternating Picture Appears   311

9-33 CAPSTAN DOES NOT ROTATE

Figures 5-20 and 7-17 show the circuits involved. This symptom assumes that the capstan does not rotate in any mode.

Start by checking that power is available as described in Sec. 9-24. Make certain that the camcorder is not in the STOP, FAST FORWARD, REWIND, or any mode where the capstan motor is supposed to be off.

With the camcorder in PLAYBACK, check for about 2.75 V at pin 8 of IC604. If present, check for about 11.5 V at pin 6 of IC604. If absent or abnormal, check the switching regulator (IC906, Q952).

If all voltages to IC604 are good, but the capstan does not rotate, suspect IC604 or the capstan motor.

It is also possible that IC901 is not producing the correct control signals to IC904. Check the truth table shown in Fig. 7-17.

If the voltage at pin 8 of IC604 is absent or abnormal, check for about 2.5 V at pin 5 of IC607 and pin 15 of IC603. If present at IC607-5, but not at IC604-8, suspect IC607, C649, and R648. If present at IC603-15, but not at IC607-5, suspect R644 and R645.

If the voltage at IC603-15 is zero, check the voltage at pins 7 and 10 of IC603. During STOP, FAST FORWARD, REWIND, and so on, pin 7 goes high, turning on the switch circuits within IC603. This pulls pin 10 of IC603 low, cutting off the capstan motor.

Also check the voltage at pins 8 and 9 of IC603. During pause, pin 8 goes high, turning on another switch circuit within IC603. This pulls pin 10 of IC603 to ground through R693 and pin 9, and shifts the d-c level at IC607.

9-34 PICTURE SWINGS HORIZONTALLY, OR NOISY PICTURE AND CLEAN PICTURE ALTERNATELY APPEAR ON THE DISPLAY IN PLAYBACK

Figures 5-15 and 5-16 show the circuits involved.

The first step in troubleshooting this symptom is to play back a known-good tape recorded on another camcorder or VCR. If the symptom is removed when the known-good tape is played, it is possible that the sync signal is not being recorded on the tape during the CAMERA (RECORD) mode.

Check for sync pulses at pins 17 and 18 of IC601. If absent at pin 17, check the composite sync signal back to IC203-5. If present at 17, but not at 18, suspect IC602. If present at 18, but not being recorded, suspect R626, C601, or the control head.

Also check that pin 11 of IC601 is low during RECORD If not, suspect IC901.

If the symptoms are the same when a known-good tape is played back, check for the 1748-Hz PWM pulses at pin 29 of IC601 during playback. If present, suspect the LPF or R636.

If the pulses are not present at IC601-29, check at pin 7 of IC601. Pin 7 should be high during record and low during playback. If not, check D604 and the record line.

With pin 7 of IC601 low (playback), check for 3.58-MHz signals of about 360 mV p-p at pin 25 of IC601. If absent, check C608 and the 3.58-MHz line.

Next check for 30-Hz PG pulses of about 4 V p-p at pin 14 of IC401. If absent, check for the 30-Hz PG pulses at pin 3 of IC610. If present at IC610-3, but not at IC601-14, suspect D615 and C617.

If the 30-Hz PG pulses are absent at IC610-3, check for 45-Hz PG pulses of about 5 V p-p at pin 12 of IC610. If absent, suspect C627 and the cylinder motor. Also check for 360 Hz cylinder FG pulses from IC602. (However, the 360-Hz pulses are probably good, if the cylinder motor is operating at the correct speed.)

If the 30-Hz PG pulses are present at IC601-14, check for pulses of about 2.7 V p-p at pin 13 of IC601. If absent or abnormal, try correcting the condition by adjustment of RT603. Also, check R615, RT603, and C618, and determine that the standby 5-V line is 5 V (during playback).

9-35 PICTURE SWINGS VERTICALLY, OR NOISY PICTURE AND CLEAN PICTURE ALTERNATELY APPEAR ON THE DISPLAY IN PLAYBACK

Figure 5-21 shows the circuits involved.

The first step in troubleshooting this symptom is to play a known-good tape recorded on another camcorder or VCR. If the symptom is removed when the known-good tape is played, switch to record mode and check for 720-Hz CFG pulses of about 2.8 V p-p at pin 6 of IC601. If absent, suspect C645.

Next check for about 2 V at pin 4 of IC601. If absent, suspect R607 and R608, and check the standby 5-V line.

Next check for about 4.4 V at pin 7 of IC601. If absent, check D604 and the record line.

Return to the PLAYBACK mode and check that pin 4 of IC601 is 0 V. Then check for 437-Hz PWM pulses of about 4.3 V p-p at pin 2 of IC601. If the pulses are present, try correcting the condition by adjustment of tracking control RV101 and the tracking preset control RT601. If this does not cure the trouble, suspect R644 and the LPF at pin 2 of IC601.

If the 437-Hz pulses are absent at IC601-4, check for pulses of about 2.8 V p-p at pin 19 of IC601. If absent or abnormal, suspect C610, RT601, and RV101, and check the 5-V line.

Sec. 9-37   Alternating Picture Appears on Screen (When Tape Is Played Back)   313

If the pulses are good at IC601-19, check for 30-Hz pulses of about 3.7 V p-p at pin 21 of IC601. If pulses are present at pins 19 and 21, but absent or abnormal at pin 2, suspect IC601.

If the pulses are absent at IC601-21, check for pulses at pin 1 of IC606. If absent, suspect the control head, C602, R627, and C603.

If pulses are present at pin 1 of IC606, check for pulses of about 0.8 V p-p at pin 3 of IC606. If absent, suspect IC606.

If the pulses are present at IC606-3, check for pulses of about 0.6 V at pin 5 of IC606. If absent, suspect C604, R630, and C615.

9-36 MANY NOISES APPEAR ON SCREEN IN PLAYBACK MODE

Figure 5-20 shows the circuits involved. This symptom assumes that excess noise appears on both the EVF and monitor TV during playback.

Start by checking adjustment of capstan speed control R605 as described in Chapter 8.

If this does not cure the problem, check for pulses of about 3.8 V p-p at pin 6 of IC603. If absent or abnormal, check for capstan FG pulses of about 20 mV p-p at pin 2 of IC603. If absent or abnormal, suspect C636, C650, or the capstan motor.

If the pulses are good at IC603-2, check for pulses of about 2.6 V p-p at pins 4 and 5 of IC603. If absent at pin 4, suspect IC603. If present at pin 4, but absent at pin 5, suspect R655 and C638.

If the pulses are good at IC603-6, check for sawtooth pulses of about 2.5 V p-p at pin 12 of IC603. If absent or abnormal, suspect C634, R689, RT605, and IC603.

Next check for a voltage of about 2.5 V at pin 13 of IC603. If absent or abnormal, suspect C635 and IC603.

Next check for a voltage of about 2.5 V at pin 15 of IC603. If absent or abnormal, suspect IC603.

If the voltage at pin 15 of IC603 is good, but there is excessive noise during playback that cannot be adjusted by RT605, suspect IC607, R645, R644, R648, C649, and IC604.

9-37 NOISY PICTURE AND CLEAN PICTURE ALTERNATELY APPEAR ON SCREEN (ONLY WHEN PRERECORDED TAPE IS PLAYED BACK)

Figures 5-15 and 6-4 show the circuits involved.

Start by playing back a prerecorded tape. First check that pin 4 of IC610 is 0 V. If pin 4 is high (5 V), check the record line.

Next check for SW 15-Hz pulses of about 2.6 V p-p at pin 19 of IC610. If absent or abnormal, check for SW 30-Hz pulses of about 4 V p-p at pin 2 of IC610. If pulses are present at pin 2, but not at pin 19, suspect IC610.

If pulses are absent at IC610-2, check for pulses at pin 15 of IC601. If absent, check for pulses of about 2.6 V p-p at pin 12 of IC601. If absent at pin 15, but present at pin 12, suspect IC601. If absent or abnormal at pin 12 of IC601, suspect C619, R614, RT602, and the standby 5-V line.

If the pulses are good at IC610-19, check for head-switching pulses of about 3.6 V p-p at pins 6 through 9 of IC610. If absent, suspect IC610.

If the pulses are good at pins 6 through 9 of IC610, but the symptom occurs, suspect IC201, IC202, and the video heads.

INDEX

A/C head adjustments, 244
A/V adapters, 19, 23
ABO circuits, 86
Adapter/charger, 204
　adjustments, 283
AFC problems (TV), 289
AGC adjustments, 252
AIC/AGC circuits, 95
AIC circuits, 130
AIC level adjustments, 251
Alignment tape, 63
Audio
　adjustments, 282
　circuits, 192
　troubleshooting, 196
Autofocus
　adjustments, 272–75
　chart, 61
　circuits, 113
Automatic white balance, 22

Backfocus
　adjustments, 255
　chart, 61
Backlight, 53
Back tension adjustments, 236
Baseband signals, 62
Battery charger, 39, 214
Beam current adjustments, 252
Bias (audio) adjustments, 282
Bias lights, 76
Black and white (B&W), 13

Brake, tape transport, 150
Burst frequency adjustments, 248
Burst level adjustments, 268

Camera circuits (MOS), 118
Camera deflection circuits, troubleshooting, 89
Camera signal circuits, troubleshooting, 109
Capstan
　adjustments, 277
　motor troubleshooting, 229
　phase, 169
　phase troubleshooting, 171
　speed, 167
　speed troubleshooting, 169
Carrier separation circuits, 100
Cassettes, 45
　cleaning and lapping, 66
　up switch, 219
CCD pickup, 11
Character generator, 43
　circuits (MOS), 137
Charts, adjustment, 61
Chroma
　adjustments, 256, 260, 266, 268, 269, 280
　circuits, 104
　circuits (MOS), 132
　encoder, 105
　troubleshooting, 180–84, 199
Cleaning, 65
Close ups, 52
Collections, camcorder, 37

Color balance, 22
Color blur or shading, 94
Color camera, basic, 15
Color chart, 61
Color difference, 17
Color image sensor, 118
Color pickup, 11
Color principles, 10
Color temperature, 11, 61
Controls, operating, 30, 33
Copy problems, 289
Copyright problems, 59
Creased tape, 241
Cylinder
  adjustments, 276
  circuits, 154–57
  lock, 219
  phase, troubleshooting, 165
  speed, troubleshooting, 159

Dark current dispersion, 93
Dark offset adjustments, 255
Dark shading adjustments, 254
Date, recording, 54
Deflection circuits, camera
  horizontal, 79
  vertical, 82
Delta measurement, 114
Dew sensor, 217
Diopter, 21
Dynamic focus, adjustments, 248, 263, 266

Earphones, 22
Encoding circuits (MOS), 135
End sensors, 217
EVF
  adjustments, 270–72
  circuits, 116
  displays, 21, 43

Fast-forward, 149
Filter
  MOS, 124
  stripe, 15
Fine edit, 50
Flutter adjustments, 269
Focus, 22, 50
  adjustments, 262–66
  automatic, circuits, 113
  dynamic, 84
Frame joints, 175
Function switch, troubleshooting, 216

Grayscale chart, 61
Guide rollers, 243

Headswitching, 7
Horizontal adjustments, 257
Horizontal aperture circuits, 98
Horizontal aperture (MOS), 128

HQ VHS circuits, 22
HV power supply, 87

Impedance roller height, 240
Indicators, operating, 30, 33
Interchange, 242, 291
Iris
  automatic, 22
  circuits (MOS), 130

Kelvin, 11

Lapping cassettes, 66
Lens, 21
  camera, 70
Light box, 61
Light meter, 61
Light source, 61
Light, principles, 10
Linearity adjustments, 257
Loading, 146
  motor, troubleshooting, 226
Logic probe, 62
Low-light adjustments, 269
Lubrication, 67
Lumanance
  adjustments, 281
  circuits, 98
  circuits (MOS), 128
  troubleshooting, 180–84
Lux, 61

Matrix (MOS), 124
Mechanical parts, 143
Mechanism-state adjustments, 247
Mechanism-state switch, 219
Microphone, 22
Mode control, troubleshooting, 232
Mode indicator, troubleshooting, 224
Mode-sense adjustments, 247
Mode-sense switch, 219
Monitor TV, 61
MOS circuits, troubleshooting, 139
MOS pickup or sensor, 118

Newvicon, 11
Noise (picture), 175
NTSC video signal, 17

ODC adjustments, 276
Optical lens, 12
Optics, camera, 70

Parabolic circuits, 83
Pattern box, 61
Pause, 22
Percival compensation circuits, 92
PG shifter adjustments, 279
Pickup, 19
Pickup tube, circuits, 85

Index

Picture noise, 175
Playback/record, 148
Playback, troubleshooting, 184
Playback (with TV), 55
Power circuits (camera), troubleshooting, 74
Power control, troubleshooting, 211
Power supply, 21, 36
Power supply, HV, 87
Preamplifier circuits, 91
Preamplifier (MOS), 124
Prevideo circuits, 92
Primary colors, 11
PWM, 204

Quadrature modulation, 19
Quick review, 22

Rainbow effect, 175
Rc/Bc separation circuits, 100
Receiver/monitor, 62
Record/playback, 148
Record/playback sequence, basic, 47
Record/review, 50
Record, troubleshooting, 180
Red/blue gain-control circuits, 103
Reel sensor, 218
Reel table, height adjustments, 235
Remote control, 54
Remote switch, 43
Rewind, 149

Safety tab switch, 219
Saticon, 11, 73
Sawtooth circuits, 83
Search, 23, 157
Sensor-drive (MOS), 122
Servo circuits, 152
Servo system, 292
Shading adjustments, 249, 264, 267
Shading/tracking circuits, 101
Siemens star chart, 61
Signal gain adjustments, 262
Signal separation adjustments, 261
Slack tape, 241
Special effects, 55
Stop, 23
Streaking adjustments, 253
Stripe filter, 15, 73
Sync generator circuits, 78
Sync generator circuits (MOS), 122
System-control, 290

Tape
  alignment, 63
  counter, 49
  counter, troubleshooting, 222
  drive and path, 8
  guide, height adjustment, 240
  transport, 143
  travel, 240

Target point, 13
Target voltage adjustments, 250
Temperature, color, 11
Tension-arm position adjustments, 235
Test equipment and setup, 247
Torque confirmation adjustments, 237
Tracking adjustments, 250, 265, 268, 280
Tripod, 60
Trouble sensor troubleshooting, 220
Troubleshooting
  adapter/charger, 205
  audio, 196
  battery ODC, 214
  camera deflection circuits, 89
  camera signal circuits, 109
  capstan motor, 229
  capstan phase, 171
  capstan speed, 169
  chroma, 180-84, 199
  cylinder, 159
  cylinder phase, 165
  function switch, 216
  loading motor, 226
  luma, 197
  luma/chroma, 180-84
  mode-control, 232
  mode indicator, 224
  MOS circuits, 139
  playback, 184
  power circuits, 74
  power control, 211
  record, 180
  symptoms, 293-95
  tape counter, 222
  trouble sensor, 220
  video, 187, 189, 192
  video input/output selection, 179

Unloading, 147

VCR adjustments, 276-82
Vertical adjustments, 257
Vertical-edge adjustments, 259
Vertical-edge correction circuits, 101
Video input/output selection troubleshooting, 179
Video processor circuits, 105
Video troubleshooting, 187, 189, 192

White balance, 11, 22, 51
  adjustments, 270
  circuits, 107
  circuits (MOS), 133
Wow and flutter, 290

Y-enhancer (MOS), 127

Zoom, 52
Zoom motor control, 112